ENERGY SECURITY

Transforming Environmental Politics and Policy

Series Editors:

Timothy Doyle
Keele University, UK and University of Adelaide, Australia

Philip Catney
Keele University, UK

The theory and practice of environmental politics and policy are rapidly emerging as key areas of intense concern in the first, third and industrializing worlds. People of diverse nationalities, religions and cultures wrestle daily with environment and development issues central to human and non-human survival on the planet Earth. Air, Water, Earth, Fire. These central elements mix together in so many ways, spinning off new constellations of issues, ideas and actions, gathering under a multitude of banners: energy security, food sovereignty, climate change, genetic modification, environmental justice and sustainability, population growth, water quality and access, air pollution, mal-distribution and over-consumption of scarce resources, the rights of the non-human, the welfare of future citizens – the list goes on.

What is much needed in green debates is for theoretical discussions to be rooted in policy outcomes and service delivery. So, while still engaging in the theoretical realm, this series also seeks to provide a 'real world' policy-making dimension. Politics and policy-making is interpreted widely here to include the territories, discourses, instruments and domains of political parties, non-governmental organizations, protest movements, corporations, international regimes, and transnational networks.

From the local to the global – and back again – this series explores environmental politics and policy within countries and cultures, researching the ways in which green issues cross North-South and East-West divides. The 'Transforming Environmental Politics and Policy' series exposes the exciting ways in which environmental politics and policy can transform political relationships, in all their forms.

Other titles in the series:

Community Gardening as Social Action
Claire Nettle

Energy, Governance and Security in Thailand and Myanmar (Burma)
A Critical Approach to Environmental Politics in the South
Adam Simpson

Energy Security in Japan
Challenges After Fukushima

VLADO VIVODA
Griffith University, Australia

LONDON AND NEW YORK

First published 2014 by Ashgate Publishing

Published 2016 by Routledge
2 Park Square, Milton Park, Abingdon, Oxfordshire OX14 4RN
711 Third Avenue, New York, NY 10017, USA

First issued in paperback 2016

Routledge is an imprint of the Taylor & Francis Group, an informa business

British Library Cataloguing in Publication Data
A catalogue record for this book is available from the British Library.

The Library of Congress has cataloged the printed edition as follows:
Vivoda, Vlado.
Energy security in Japan : challenges after Fukushima / by Vlado Vivoda.
pages cm. – (Transforming environmental politics and policy)
Includes bibliographical references and index.
ISBN 978-1-4094-5530-1 (hardback)
1. Energy policy–Japan. 2. National security–Japan.
3. Environmental policy–Japan. 4. Fukushima Nuclear Disaster, Japan, 2011. 5. Japan–Politics and government. 6. Japan–Environmental conditions. I. Title.
HD9502.J32V59 2014
333.790952–dc23

2013034232

ISBN 13: 978-1-138-27031-2 (pbk)
ISBN 13: 978-1-4094-5530-1 (hbk)

Contents

List of Figures and Tables

Figures

Tables

Series Editors' Preface

The beginnings of this series emerged at Keele University in a collaboration between Tim Doyle and Phil Catney. Since the late 1970s, Keele has been renowned across the globe as one of the leading universities engaged in teaching and research into the politics and international relations of the environment.

Our initial conversations with Ashgate were around two objectives. First, we wanted to transform the rather narrow, dominant conceptions of environmental politics and policy – particularly in the global North – by opening it up to include issues more central to traditional human politics and policy-making. Gone are the days when the 'environment' is something that people engage in (and with) as some kind of 'luxury' pursuit, when and if they have the time and the resources to do it. Nowadays, environmental issues – both in the North and the South – are front and centre. Secondly, we strongly felt that much needed in environmental debates was for theoretical discussions to be rooted in policy outcomes and service delivery. In short, the series would look at the exciting ways in which environmental politics and policy can transform governance, in all its forms.

Timothy Doyle, Keele University, UK and University of Adelaide, Australia,
and Philip Catney, Keele University, UK

Acknowledgements

The inspiration for this book stems from an energy security study tour organized by Japan Foundation in May 2011. Having witnessed, first hand, the effects of the Fukushima disaster on Japan's society, economy and energy security shortly after the events of 11 March 2011, motivated me to explore Japan's energy challenges. The initial intellectual and analytical basis for the book was an article ('Japan's Energy Security Predicament post-Fukushima') published in *Energy Policy* (vol. 46, 135–43) in 2012. In this book I have extended my analysis to include more detailed and up-to-date assessment of Japan's energy challenges after the disaster.

I am appreciative for the research funding provided by the Australian Research Council (DE120101090) and the National Research Foundation of Korea (NRF 2012S1A3A2033350), which have supported elements of this work. I would also like to thank Japan Foundation for funding the energy security study tour of Japan in May 2011.

Some of the findings of this book are supported by interviews conducted in Japan in January 2013 and I would like to thank the respondents for their time and input. Notwithstanding their help, all errors and conclusions are entirely my responsibility. I would also like to thank my colleagues at Griffith University for feedback provided at seminars where I presented various chapter drafts; the anonymous reviewer for constructive and very helpful feedback on an earlier draft; Geordan Graetz for his assistance with some of the research for the chapter on nuclear energy and with the index; and Darko Vivoda for his help with graphs. Finally, I wish to extend a note of appreciation to my wife, Alice, for her love and support throughout the course of this book project.

Vlado Vivoda

List of Abbreviations

ANRE	Agency for Natural Resources and Energy (Japan)
AOC	Arabia Oil Company
APC	Atomic Power Company
BEP	Basic Energy Plan
bpd	barrels per day
CCS	carbon capture and storage
cif	cost–insurance–freight
CNPC	China National Petroleum Corporation
DoE	Department of Energy (US)
DPJ	Democratic Party of Japan
Enecan	Energy and Environmental Council
ESPO	East Siberia Pacific Ocean (pipeline)
FDI	foreign direct investment
FEPC	Federation of Electric Power Companies of Japan
FIT	feed-in-tariff
FOB	freight on board
FTA	free trade agreement
GDP	gross domestic product
GHG	greenhouse gas
GSP	government selling price
GTL	gas-to-liquids
GW	gigawatt
IAEA	International Atomic Energy Agency
IEA	International Energy Agency
IEEJ	Institute of Energy Economics, Japan
IGCC	integrated gasification combined cycle
IPPs	Independent Power Producers
J-EXIM	Export–Import Bank of Japan
J-POWER	Electric Power Development Company
JAEA	Japan Atomic Energy Agency
JAEC	Japan Atomic Energy Commission
JAERI	Japan Atomic Energy Research Institute
JAPEX	Japan Petroleum Exploration Company
JAURD	Japan Australia Uranium Resources Development Co. Ltd.
JBIC	Japan Bank for International Cooperation
JCC	Japan Crude Cocktail
JCOAL	Japan Coal Energy Center

JGDC	Japan Geothermal Developers' Council
JNC	Japan Nuclear Cycle Development Institute
JNOC	Japan National Oil Corporation
JOGMEC	Japan Oil, Gas and Metals National Corporation
JPDC	Japan Petroleum Development Corporation
JPEA	Japan Photovoltaic Energy Association
KEPCO	Kansai Electric Power Company
kl	kilo litres
KOGAS	Korea Gas Corporation
kW	kilowatt
kWh	kilowatt-hours
LDP	Liberal Democratic Party
LNG	liquefied natural gas
LPG	liquid petroleum gas
LWR	light water reactor
METI	Ministry of Economy, Trade and Industry (Japan)
MEXT	Ministry of Education, Culture, Sports, Science & Technology (Japan)
MITI	Ministry of International Trade and Industry (Japan)
MoE	Ministry of Environment (Japan)
MoF	Ministry of Finance (Japan)
MoFA	Ministry of Foreign Affairs (Japan)
MOX	mixed-oxide
mtoe	million tons of oil equivalent
mtpa	million tons per annum
MW	megawatt
NEDO	New Energy Development Organization/New Energy and Industrial Technology Development Organization
NIMBY	not-in-my-backyard
NISA	Nuclear and Industrial Safety Agency
NNES	New National Energy Strategy
NOSODECO	North Sumatra Oil Development Cooperation Co. Ltd.
NRA	Nuclear Regulation Authority
NSC	Nuclear Safety Commission
NSIC	Nuclear Safety Investigation Committee
OAPEC	Organization of Arab Petroleum Exporting Countries
ODA	official development assistance
OECD	Organization for Economic Cooperation and Development
OPEC	Organization of Petroleum Exporting Countries
PNC	Power Reactor and Nuclear Fuel Development Corporation
PV	photovoltaic
RPS	Renewable Portfolio Standard
SLOCs	sea lines of communication
SPA	sales and purchase agreement

STA	Science and Technology Agency (Japan)
TEP	Tōhoku Electric Power Company
TEPCO	Tokyo Electric Power Company
tU	tonnes of uranium
TWh	terawatt-hours

Introduction
The Fukushima Disaster and Its Aftermath

The 11 March 2011 Tōhoku earthquake and tsunami caused 15,879 deaths, 6,130 injuries, with 2,698 people missing across 20 Japanese prefectures. In addition, 128,884 buildings totally collapsed, with a further 268,905 buildings 'half collapsed' (National Police Agency of Japan 2013). The earthquake and tsunami also caused extensive and severe structural damage in north-eastern Japan, including heavy damage to roads and railways as well as fires in many areas and a dam collapse. The tsunami caused a number of nuclear accidents, primarily the level 7 meltdowns at three reactors in Tokyo Electric Power Company's (TEPCO's) Fukushima Daiichi Nuclear Power Plant complex and the associated evacuation zones affecting hundreds of thousands of residents. Many electrical generators were taken down and at least three nuclear reactors suffered explosions due to hydrogen gas that had built up within their outer containment buildings after cooling system failure. Residents within a 20 km radius of the Fukushima Daiichi Nuclear Power Plant and a 10 km radius of the Fukushima Daini Nuclear Power Plant were evacuated. The World Bank's estimated economic cost was US$235 billion (Kim 2011). In the immediate aftermath of the disaster, Japanese Prime Minister Naoto Kan said, 'In the 65 years after the end of World War II, this is the toughest and the most difficult crisis for Japan' (cited in CNN 2011). Andrew DeWit (2011) described the disaster as the costliest natural catastrophe in human history that will be matched by history's most expensive rebuild.

As an immediate effect of the disaster on energy supplies, 4.4 million households served by Tōhoku Electric Power (TEP) in north-eastern Japan were left without electricity (NPR 2011). Several nuclear and thermal power plants went offline after the earthquake. As most of the thermal power plants in eastern Japan are located on the Pacific coast, damage was severe and widespread. Many were flooded by the tsunami, which inundated them with soil and debris and caused severe damage to turbines and other equipment. Oil refineries suffered damage, coal unloading facilities collapsed and coal carriers sank, thus affecting fuel supplies. Six refineries were shut down and there was damage to transportation routes, storage facilities and petrol stations (Petroleum Association of Japan 2012: 67). As a consequence, there was localized shortage of petroleum supply around Japan in the immediate aftermath of the disaster. Similarly, town gas supply was disrupted in devastated areas, with supply for 420,000–460,000 users stopped and Minato LNG importing terminal in Sendai was shut down and severely damaged (Yamashita 2012, Miyamoto et al. 2012). Moreover, close to one-third of thermal power capacity in eastern Japan was affected. By type of fuel, around one-third of total oil-fired thermal capacity was affected, while three-quarters of coal-fired thermal capacity in eastern Japan was lost. The combination of reduced dependence on coal as a baseload power source and shutdown of nuclear power plants had a major impact on electricity supply capacity (Miyamoto et al. 2012: 4, Sagawa 2012).

As a consequence, TEP was not able to provide the Kanto region with additional power, because its power plants were damaged. Kansai Electric Power Company (KEPCO) was unable to share electricity, because its system operates at 60Hz, whereas TEPCO and TEP operate their systems at 50Hz. Two substations, one in Shizuoka Prefecture and one in Nagano Prefecture, were able to convert between frequencies and transfer electricity from Kansai to Kanto and Tōhoku, but their capacity was limited to around 1 GW. Rolling blackouts began on 14 March due to power shortages caused by the earthquake. TEPCO could provide only three-quarters of its usual electricity supply. This was because 40 per cent of the electricity used in the greater Tokyo area had been supplied by nuclear reactors in the Niigata and Fukushima prefectures. The reactors at the Fukushima Daiichi and Fukushima Daini plants were automatically taken offline when the first earthquake occurred and have sustained major damage related to the earthquake and subsequent tsunami. Rolling blackouts of approximately three hours were experienced throughout April and May 2011 while TEPCO scrambled to find a temporary power solution.

The Japanese government and electric utilities took several steps to ensure power supply meets demand following the Fukushima crisis. As an immediate response to the disaster, and with the aim of increasing thermal power supply, the Japanese government allowed for an exemption from the *Environmental Impact Assessment Act* for thermal power plant expansion and approved the delay of periodic inspections of thermal power plants. The compulsory oil stockpile regulation for private companies was reduced to 45 days to allow for more petroleum to be available to the market. At the same time, the electric utilities responded by restoring some of the damaged thermal power plants, restarting long-idled thermal power plants, installing new emergency power generators (such as gas turbines) and increasing electricity transfer among interconnected regions (Yamashita 2012). In addition, the government promoted power restraints for consumers in the disaster-affected areas throughout 2011, invoking a 15 per cent power reduction on all consumer groups. The Energy and Environment Council concluded that the government would need to request voluntary power saving efforts of 10 per cent and 5 per cent, respectively, from end users of KEPCO and Kyushu Electric Power Company during the summer of 2012. Also, the government requested that four western service areas with surplus capacity to cut electricity consumption by five per cent in order to transfer power to the north eastern power areas with electricity deficits (EIA 2012b). In addition to cuts in consumption enforced by rolling power cuts and restrictions on electricity use, the public and private sectors together ran campaigns to encourage voluntary conservation (*mottainai* or 'don't waste'), resulting in reduced energy (and especially electricity) consumption nationwide (Miyamoto et al. 2012: 24–25).

As a result, Japan's overall electricity supply was able to meet the electricity demand of the 2011 summer peak and the 2011–2012 winter peak by significantly reducing demand to less than that of previous peaks – a result of the nationwide comprehensive electricity conservation measures (Hayashi and Hughes 2013). For example, in the summer of 2011 stringent energy conservation measures were applied leading to a 12 per cent reduction in power consumption (relative to 2010) in August and, more significantly, a reduction in peak demand reaching 18 per cent, exceeding the government target of 15 per cent. Peak electricity demand for TEPCO and TEP was 16.5 per cent and 19.8 per cent, respectively, lower in 2012 than in 2011, helping to avoid summer outages (IEEJ 2011c: 1).

While the availability of thermal power and hydroelectricity improved quickly after the earthquake and tsunami, the availability of nuclear power deteriorated with the indirect impacts of the accident. Between the 2011 earthquake and May 2012, Japan lost all of its nuclear capacity due to scheduled maintenance and the challenge facilities face in gaining government approvals to return reactors to operation. Two reactors were restarted in Ōi in July 2012 to tackle looming electricity shortages in the Kansai region during summer, but have been shut again in September 2013 for routine maintenance checks. With nuclear stations providing 25–30 per cent of Japan's electricity before the Fukushima disaster, utilities have been forced to rely more on oil- and gas-fired power plants to make up the difference. Electricity from thermal generation increased gradually to a level greater than before the earthquake, a consequence of the actions taken by the government and electricity suppliers to complement the decline of nuclear power. Consequently, the Fukushima nuclear accident changed Japan's electricity supply and overall energy structure dramatically. The shutdown of nuclear power plants led to a sharp rise in consumption of fossil fuels in the power generation sector, increasing demand for both LNG and low-sulphur fuel oil and crude oil in order to supplant lost nuclear power generation. The share of thermal generation as a share of total generation increased from 63 per cent in 2010 to 74 per cent in 2011 and to 88 per cent in the first ten months of 2012, the highest on record (IEA 2013a: 24). At the same time, the share of fossil fuels in Japan's energy mix increased from 81.9 per cent in 2010 to 93.6 per cent in 2012 (BP 2013).

The increased use of thermal plants to make up for the loss of nuclear output caused higher fuel import costs, borne by Japanese consumers and industries, leading to a first trade deficit since 1980. The 2011 and 2012 trade deficits stood at ¥2.56 trillion and ¥6.93 trillion, respectively. In both years, trade deficits were mainly caused by an increase in the value of fossil fuel imports. With Fukushima and other nuclear plants offline, the value of Japan's mineral fuel imports increased from ¥17.4 trillion in 2010 to ¥21.8 trillion in 2011 and ¥24.1 trillion in 2012 (MoF 2013). Japan's Ministry of Economy, Trade and Industry (METI) estimate that electricity costs would need to increase up to 20 per cent while the nuclear plants remain idle (World Nuclear Association 2012). As a consequence of the nuclear shutdown and increased cost of energy imports, corporate customers in and around Tokyo have been paying up to 18 per cent more for their electricity beginning April 2012 (Soble 2012). In 2012, a regular household's electricity bill was predicted to increase by 18 per cent and the rate for industrial consumers by 36 per cent on average due to rise in fuel costs (IEEJ 2011a). Residential and industrial electricity prices are already considerably higher in Japan than in most G-20 economies (IEA 2011a).

Over the past decade, high fuel costs have not only handicapped the ability of Japanese industry to compete in a globalizing market but have also impaired Japanese consumers' purchasing power and thus Japan's economic growth potential (Yokobori 2005: 311). The deeper the cut in nuclear use for power generation, the larger the negative impact on GDP (Itakura 2011). The higher fuel cost since the Fukushima disaster has reduced Japan's GDP by an estimated 0.9 per cent in 2011 and 1.3 per cent in 2012. The growing dependence of the Japanese economy on imported energy severely affects the potential for unhindered economic growth. The economic impact on Japan without nuclear energy is profound and the hollowing out of manufacturing industries will escalate as manufacturers can no longer protect jobs in Japan and instead move to neighbouring countries where costs are lower (Hosoe 2012b: 53). Japan's major financial newspaper, the *Nikkei Shinbun*, published a series of surveys showing that many corporations plan to relocate their manufacturing to offshore locations – including India, China and Malaysia – if the Japanese government cannot create a plan to ensure stability in the electricity supply over the next three years. One Japanese business analyst argued that 'if we completely abandon nuclear power generation … I think most industries would lose competitiveness and go out of Japan'.

Many observers have underscored the fact that corporations dislike uncertainty, and uncertainty about disruptions in Japan's power supply (or a spike in electricity costs) has made many Japanese firms deeply anxious. Given the economic difficulties the nation has faced over the past two decades, these new economic threats are being taken very seriously (Aldrich 2012: 7). The Institute of Energy Economics, Japan (IEEJ), Chairman and CEO Masakazu Toyoda commented: 'the zero nuclear policy could cause the hollowing out and collapse of the Japanese economy' (IEEJ 2012g: 3). Akio Mimura, Chairman of the Basic Energy Policy Subcommittee argued in September 2012 that 'nuclear energy should not be abandoned. Abandoning what we have now while the future remains uncertain will greatly threaten our energy security and energy diplomacy. The irreversible consequences of pursuing the zero nuclear policy should be explained thoroughly to the public' (IEEJ 2012g: 3).

There are also severe consequences for Japan's environmental policy following a reduction in nuclear output. Japan's CO_2 emissions increased by 2.1 per cent in 2011 and, with most nuclear reactors shut in 2012, CO_2 emissions increased by further 6.7 per cent (BP 2013). If there were no restrictions on resuming operations in Japan's nuclear reactors, a 5.3 per cent drop in CO_2 emissions was predicted for 2012 (IEEJ 2012a). Before Fukushima, nuclear power reduced Japan's CO_2 emissions by 14 per cent per year (EIA 2011a, Nakano 2011). Increased emissions make it virtually impossible for Japan to reach the Kyoto Protocol 2020 target of reducing CO_2 emissions by 25 per cent of 1990 levels. Within a year from the disaster, Japanese leaders have been frank in dismissing any hopes of meeting Japan's climate change targets (*World Nuclear News* 2012). In November 2013, Japan has announced it is significantly reducing its greenhouse gas reduction target. It aims to achieve a 3.8 per cent cut in carbon dioxide emissions by 2020 versus 2005 levels. The new target amounts to a 3.1 per cent increase from 1990 levels, a sharp reversal from the 25 per cent reduction target. It is worth noting that Tokyo has been a world leader in pushing for greater use of very low carbon emission sources.

The Energy Challenge

From a long-term viewpoint, Japan faces a very serious energy security problem, pertaining to overseas supplies, energy cost, the status of nuclear power, greenhouse gas emissions and growth in renewable energy. Japan is the world's fifth largest energy consumer and a resource-poor country, which imports close to all of its fossil fuel requirements. Japan's energy challenge associated with increased demand for imported fossil fuels is exacerbated by the fact that major economies in the Asia–Pacific region are competing for supplies of fossil fuels and particularly oil. Energy supply in Japan is 90–95 per cent dependent on overseas imports (IEA 2012), which when coupled with high and volatile energy prices in recent years make Japan particularly vulnerable to supply shocks and price volatility. This is exacerbated by the shift in fuel portfolio for power generation towards fossil fuels as some of Japan's nuclear reactors remain permanently offline after the Fukushima disaster.

If we are to define energy security as the availability of energy at all times in various forms, in sufficient quantities and at affordable prices, without unacceptable or irreversible impact on the economy and the environment (UNDP 2004), Japan is facing a serious predicament and a dilemma regarding the direction of its future energy policy. As a consequence of the Fukushima disaster, the Japanese people are paying more for energy, the supply of which is less secure. Moreover, the higher cost of the energy mix, which is heavier on fossil fuels, has an adverse effect on both the economy and the environment. Consequently, the nuclear crisis poses a serious challenge to the nation's economy and its energy security in terms of affordability, supply security, safety and the environment.

Large demand for energy and high import dependence has made energy security as one of the priorities of any government in Tokyo, particularly since the two oil crises in the 1970s. The 1973 and 1979 oil crises caused the Japanese economy to record negative growth rates for the first time in its post-war history. Their impact on the lives of ordinary Japanese remains deeply etched on people's minds. As a result, the Japanese government adopted policies aimed at improving energy efficiency and reducing the demand for oil. These policies have resulted in unprecedented success. Consequently, Japan is now the most energy-efficient country in the world (*The Economist* 2011). In addition, Japan's oil demand dropped from 5.4 million bpd in 1979 to 4.4 million bpd prior to Fukushima, due to vehicle efficiency gains and conversion to other electricity sources. The share of oil in total energy consumption has declined from about 72 per cent in 1979 to 40 per cent prior to the Fukushima disaster (BP 2013).

Today, after more than three decades, energy security is once again at the centre of attention among Japanese policy-makers and the general public. Faced with suspended lifeline networks owing to energy supply disruptions for the first time in over three decades since the second oil crisis, the Japanese people have re-awakened to the importance of energy security. However, unlike in the 1970s, when the focus was on affordability and security of oil supplies, the current challenge is multidimensional. While the renewed interest in energy security issues was triggered by record oil prices in 2008, it was brought to the forefront of public debate in the aftermath of the earthquake and tsunami, which caused a nuclear catastrophe in TEPCO's Fukushima Daiichi nuclear power plant.

Consequently, largely absent since the two oil crises in the 1970s, the energy debate in Japan has been revived in the aftermath of the Fukushima disaster (Calder 2013, Hayashi and Hughes 2013, Huenteler et al. 2012, McLellan et al. 2013, Moe 2012, Shadrina 2012, Tanaka 2013, Vivoda 2012). The disaster and the ensuing nuclear shutdown have created a profound national energy security crisis. Japan's energy policy has to address challenges related to the future availability of diverse energy sources, increasing cost of fuels, nuclear safety and adverse impact of its energy and power demand trajectory on the economy and the environment. Japan is the key case not only because of pressure on the existing energy system caused by the Fukushima disaster, but it also has an established track record of reacting to energy crises in the 1970s. Against the backdrop of energy policy uncertainty in Japan, this book evaluates Japan's current energy security situation and places future energy policy options in the appropriate context. As such, the book contributes to the literature on Japan's energy security and policy.

This book is timely not only because much of the existing literature on Japan's energy security pre-dates Fukushima, but also because the scale of Japan's energy challenges require urgent scrutiny in the aftermath of the disaster. The shutdown of Japan's nuclear power program adds to an array of significant existing challenges. The collateral impact of potentially abandoning this power generation program is extensive. Duffield and Woodall (2011) have analysed Japan's 2010 Basic Energy Plan (BEP) and argued that even prior to Fukushima, achievement of many targets was likely to be challenging. This is exacerbated in the aftermath of the Fukushima disaster (Vivoda 2012). In many ways, previous energy security thinking needs to be reassessed in lieu of a changed post-Fukushima environment. As one of the world's leading industrialized nations and a major importer of fuels, the choice of future energy paths by Japan will have a significant influence on the energy security of the world as a whole and of the Northeast Asia region in particular (Takase and Suzuki 2011: 6731). Consequently, the book explores change and continuity in Japan's energy security challenges following the Fukushima disaster.

Challenges were carefully selected based on the review of scholarly literature, media reports and responses from interviews conducted with Japan energy experts in January 2013. Japan's energy challenges surveyed in this book are by no means exclusive. However, they provide for a sufficient scope, which highlights their multidimensional nature. The aim of this book is not to prescribe a policy direction for Japan that will be a 'silver bullet' which would eliminate all of Japan's energy issues. Instead, the book analyses various challenges for Japan and offers suggestions on how the negative effect of these challenges on Japan's energy security may be reduced.

The book also contributes to the literature on responses to crises. A regularly invoked interpretation of policy change has divided history into 'normal periods' (or institutional stasis) and 'critical junctures', during which major change is possible (Gorges 2001, Hogan and Doyle 2007). Relatively lengthy periods of institutional stasis are punctured periodically by intense and cathartic bouts of crisis, leading to institutional change (Krasner 1984). Crises or exogenous shocks are often cited as explanations for policy change (Greener 2001, Golob 2003), as their existence highlights a failing within existing policies due to their implication in, or inability to right, the emergent situation (Levy 1994). Crises expose decision-makers to criticism and demands for more effective action (Walsh 2006), resulting in policy change. Crises unleash short bouts of intense ideational contestation in which agents struggle to provide compelling and convincing diagnoses of the pathologies afflicting the old regime/policy paradigm and the reforms appropriate to the resolution of the crisis (Blyth 2002).

However, exogenous crises do not always result in structural policy and institutional change. The introduction of new ideas into the policy environment and their transformation into policy, often takes place due to the activities of 'entrepreneurial networks' of policy entrepreneurs, with political entrepreneurs at their head. Constituents, such as policy and political entrepreneurs, generate and institutionalize emergent policy ideas (Orren and Skowronek 1994). Walsh (2006) argues that policy change is most likely to occur when an alternative policy idea can explain past failures and secure the support of powerful constituents. In order for policy entrepreneurs to challenge existing arrangements, a crisis and policy failure must be identified and widely perceived (Hay 1999). Agents must diagnose and impose on others, their notion of a crisis before collective action to resolve the resultant uncertainty can be taken (Blyth 2002). Agents shape 'the terms of political debate: they frame issues, define problems and influence agendas' (Sheingate 2003: 188). They ultimately initiate a debate concerning extant ideational orthodoxy.

Consequently, exclusive reliance on exogenous shocks to account for policy change is too simplistic and often fails to explain the absence of change in the wake of a crisis. Therefore, at a time of crisis, it is important to consider both exogenous explanations, such as the Fukushima disaster and, endogenous explanations, such as institutional sources of policy change in terms of idea generation and idea advocacy, to explain the potential for policy change (Hogan and Feeney 2012).

The Fukushima disaster is certainly a critical juncture after which major energy policy and institutional change in Japan is possible. In the aftermath of the Fukushima disaster, one idea that has gained traction is that the ensuing energy crisis is likely to induce a corresponding change in the institutions that govern Japan's energy industry. There have been calls for an overhaul in Japan's energy governance and policy and for a move away from nuclear power to renewable energy. However, it remains unclear whether the agents for change hold sufficient agency to effect structural change in Japan's energy policy and institutions that govern Japan's energy. Against this backdrop, this book answers a series of questions that relate to capacity and desirability for post-crisis reform in Japan's institutional arrangements that govern energy and in energy policy itself. Are agents for change in Japan's energy policy and institutions that govern energy sufficiently influential to inject new ideas into the political environment? Is Japan's new energy strategy going to be based upon a new paradigm and not present a mere improvement of the old centralized system? Will Japan undergo a major overhaul of institutions that govern energy? Is major energy policy change possible or desirable?

Book Structure

While recognizing the significance of the Fukushima disaster, Japan's energy future is path dependent. It is embedded in a specific political, economic and social context. Chapter 1 examines the three main sources of path dependency – interests, institutions and ideas – which affect and/or constrain Japan's future energy policy choices. More specifically, whose interests are being served by Japan's energy policy? What institutional arrangements underpin the operation of energy market mechanisms in Japan? How have these institutional structures evolved over time? Where do the ideas based on which energy policy is formulated come from and how do they become influential? Chapter 1 illustrates that Japan's energy policy-making process has been dominated by the iron triangle of bureaucrats, industry groups and politicians. These vested interests, centred on METI, the *Keidanren*, *Denjiren* and the LDP, have remained in control of Japan's policy-making apparatus for at least four decades. The predominant view among these vested interests is that the Japanese government has to maintain an interventionist role in the functioning of the energy markets, as Japan – as a resource poor country – is vulnerable to both external and internal energy shocks. Chapter 1 also argues that energy transitions are slow and protracted affairs and are another source of path dependency. More specifically, the slow nature of energy transitions serves as a major structural constraint to quick shifts in Japan's future energy mix. Against this backdrop, this chapter offers a brief account of the evolution of Japan's energy supply and demand balance, arguing that a rapid move away from fossil fuels is unlikely.

Most of Japan's energy security challenges are not new phenomena and have existed for decades. Consequently, there is historical contingency related to most of these challenges. In order for the book to proceed with effective analysis and prescriptive advice in latter chapters, it is essential to place Japan's energy security challenges in the appropriate theoretical and empirical context and to set the scene for subsequent chapters. Thus, Chapter 2 examines the historical evolution of Japan's energy policy and energy security strategy. The analysis of the evolution of Japan's energy policy in the aftermath of the 1970s oil crises also speaks to Japan's past experience with responding to external shocks. It is important to note that each of the following chapters (3–8) offers a more detailed evolution of energy policy and historical supply and demand evolution pertaining to each particular energy source in Japan.

The public and scholarly debate following the Fukushima disaster has centred on the role of renewable energy and nuclear power in Japan's future energy policy; and on the practicalities of making regulatory oversight of Japan's nuclear industry more transparent and accountable. While this debate is important and this book will contribute to it, it is important to acknowledge that both renewable energy and nuclear power are minor contributors to Japan's energy supply and that fossil fuels dominate and will continue to do so for at least the next two decades. Following the Fukushima disaster, Japan has been importing much of its oil and natural gas at a premium cost dictated by an uncertain global energy market. This raises traditional energy security concerns, especially within the current geopolitical context. In fact, traditional energy security concerns remain valid post-Fukushima and have arguably become more acute since the disaster. Even before the earthquake, there was recognition that the nuclear energy expansion would not save Japan from oil dependency in the transportation sector (Barrett 2011). Japan's energy situation begins and ends with structural constraints. Because Japan is a resource-poor industrial giant, it imports much of its primary energy supply (Vivoda and Manicom 2011). The lessons that Japan has learnt from the 1970s oil crises attest to the dangers of increased reliance on imported fossil fuels, which remain today. In fact, they are exacerbated by the increased imports of oil and natural gas to fuel thermal-fired power plants; zero-sum competition for oil and gas with China; and the US pressure to reduce oil imports from Iran. The Arab Spring, which has led to turmoil and volatility in global energy markets, has accompanied a spike in oil and gas prices arising from increased tensions between Iran and the United States and the European Union. Recent sanctions against Iran cast a long shadow on the future free flow of oil and gas through the Strait of Hormuz. Iranian oil exports have dropped from the normal level of 2.5 million bpd to 1 million bpd due to the economic sanctions and the risk of travelling through the Strait of Hormuz has increased (IEEJ 2012i: 3).

Prior to the Fukushima disaster, it was suggested that Japan's energy security has become ever more vulnerable, influenced by the fast-changing political landscape in Northeast Asia, (Jain 2007: 28). Japan's energy security challenges were predicted to grow as its relative size in Asia's energy balances declines (Evans 2006: 2). Japan's immediate need to replace its nuclear power production capacity takes place in the context of China's rise as the world's leading energy consumer. China often competes for energy resource imports on a state-to-state basis and wields considerable influence as the world's largest market. The higher cost and limited availability of thermal fuels may be the main sources of friction. While the United States and Japan may stand together on many issues, American foreign policy with Iran is one that will present specific difficulties for Japan, as it strives to secure resources for a clear and urgent need for reliable oil and gas imports. While the region does offer considerable energy resources, political uncertainty prevails and the question remains: How will Japan meet all of its energy needs?

Many of Japan's energy security challenges were present prior to the Fukushima disaster and some of these challenges were exacerbated by the disaster. This is acknowledged throughout this book, as the chapters that examine various energy sources analyse the situation both prior to and in the aftermath of the disaster. Each chapter provides an overview of the major challenges for Japan pertaining to various energy sources. Although Japan's oil demand has dropped since the 1970s oil crises, oil remains the most important energy source for Japan as it is unrivalled in the transportation sector and still heavily used in the petrochemical sector. Consequently, Chapter 3 examines Japan's major challenges related to the supply of and demand for crude oil and petroleum products. Besides the transportation sector's overreliance on oil, other challenges include continued overreliance on the Middle East, continued inefficiency of Japanese oil companies, dubious viability of Japan's equity (or self-produced) oil policy and increased Asian competition for overseas oil supplies. Unlike in subsequent chapters on other energy sources, much attention in Chapter 3 is given to the evolution of Japan's oil policy. This is a conscious choice given that oil has been the most securitized energy source in Japan over the past four decades and energy security strategy has revolved around security of oil supplies.

Japan is the pioneer in the global LNG trade and remains the world's largest importer. In recent years, buoyed by a large regional increase in LNG demand, LNG prices in the Asia–Pacific region have been considerably higher than in other regions. With increasing demand for LNG in Japan since the Fukushima disaster, Japan has been faced with increased costs for imported LNG. Chapter 4 evaluates Japan's major challenges associated with natural gas supply and demand. The major issues under scrutiny include the adequacy of domestic natural gas infrastructure in order to meet demand growth, LNG pricing and continued reliance on oil indexation and the lack of regional cooperation. Most importantly, how can Japan break away from the established LNG pricing formula in the Asia–Pacific region?

Historically, the main attraction of coal has been its plentiful supply in the Asia–Pacific region and close proximity of coal-rich countries to Japan. With Japan's commitment to greenhouse gas (GHG) emissions abatement since the Kyoto protocol was signed in 1997, the government has been committed to reducing Japan's reliance on coal, the most emissions-intensive fossil fuel. At the same time, coal remains the mainstay in Japan's energy supply mix, maintaining its status as the second largest energy source. The fuel has been particularly useful in the aftermath of the Fukushima disaster in order to replace lost nuclear power and reduce costs of LNG imports. Australian thermal coal has been crucial in this context, with Japan relying on Australia for 70 per cent of its imports. Consequently, Chapter 5 examines major challenges related to Japan's coal use. How can coal continue to play an important role in the future, while CO_2 emissions from its use are reduced? Are there dangers in overreliance on Australian thermal coal imports?

Prior to the Fukushima disaster, nuclear power played an important role in Japan's energy mix, accounting for 25–30 per cent of electricity supply. Nuclear power was considered relatively safe, cheap and a domestic source of energy. However, following the Fukushima disaster, many in Japan have called for a nuclear phase-out, with public opinion strongly opposed to continued use of nuclear energy. Regulatory capture has been cited as the main culprit for lax safety procedures that have indirectly lead to the nuclear catastrophe at Fukushima. At the same time, the iron triangle of government bureaucrats, politicians and industry remain committed to the continued reliance on nuclear energy. Against this backdrop, Chapter 6 discusses the major challenges for Japan related to nuclear energy. The focus is on the future of nuclear power in Japan, regulatory reforms and on disconnect between government policy and public opinion. Will Japan's new nuclear regulator be independent from industry capture? Can the government and industry regain public trust in nuclear power? What is the future for nuclear power in Japan?

Japan played a major global role in the development of solar energy from the late 1990s until several years ago. What has been lacking, however, is any substantial policy support for raising the share of renewables in Japan's energy mix. However, following the Fukushima disaster, the improved and generous feed-in-tariff (FIT) has been adopted in order to increase the share of renewable energy in Japan's electricity supply. Even if the high cost issue can be overcome, a number of major challenges stand on the way to a greater penetration of renewable energy in Japan and they are the focus of Chapter 7. The major challenges discussed in the chapter include institutional bias towards solar power, the potential of the new FIT and grassroots level support for renewable energy post-Fukushima to drive major penetration of renewable energy in the future and a series of structural issues pertaining to various renewable energy sources that may inhibit their growth in Japan's energy mix.

Most of the challenges discussed in Chapters 3–7 affect Japan's electricity supply. This is particularly the case with regards to nuclear power and renewable energy. Chapter 8 explains how these challenges interact with the unique structure of Japan's electricity market, dominated by ten regional electric utility monopolies. Revisions of the *Electric Power Law* of between 1995 and 1999 permitted new entrants, weakened central control of pricing and allowed more competition on the retail side. However, these piecemeal reforms have not resulted in lower electricity prices and have not challenged regional dominance of electric utilities. The utilities remain in control of both electricity generation and transmission, which presents an enormous obstacle to potential new entrants. At the same time, the electric utilities have been affected by the nuclear power shutdown and have found themselves in a difficult financial position as they are bearing the increased cost of energy imports to fuel their thermal power plants. The utilities, as part of the vested interest structure, remain committed to nuclear power and are also largely opposed to an increased penetration of renewable energy due to high costs. Finally, Japan's electricity grid is fragmented and there is limited capacity to transfer electricity between regions and between eastern and western Japan, which was a major issue in the aftermath of the Fukushima disaster. Consequently, with the discussion centred on the role of electric utilities, this chapter will evaluate some of the major challenges in Japan's electricity markets.

The conclusion summarizes Japan's major energy security challenges and the adequacy of current energy strategy and institutional arrangements that govern Japan's energy sector for tackling the existing challenges. The conclusion suggests various policies, which if implemented, may improve Japan's energy security in the future. The main argument put forward is that structural change in Japan's energy policy-making apparatus and energy strategy is unlikely. The agency of actors that support major changes in Japan in the aftermath of the Fukushima disaster is negligible when compared to the power of vested interests. As a consequence, we are likely to witness piecemeal reforms to energy governance mechanisms and energy policy, with Japan's return to nuclear power expected throughout late 2013 and 2014.

This book is a case study of Japan's energy challenges after the Fukushima disaster and the ensuing energy crisis. The data is derived from literature studies and from a set of semi-structured interviews with mainly Japan-based respondents, conducted in 2013. The respondents were selected both for their expertise and in order to bring in a broad range of viewpoints, from government and industrial interests, Japan-based foreign nationals with energy-related experience and from scholars and experts specializing in different areas of the energy field and working in universities or institutes. All agreed to being listed as respondents (see List of Respondents). Some of the respondents preferred not to be directly cited in the text. Mapping the Japanese energy structure without conducting interviews would not have been possible. Given that in Japan, much of the action happens behind the scenes, the interviews provided valuable flesh to the existing theoretical and empirical bones.

Chapter 1
Interests, Institutions and Ideas

Introduction

Path dependence can be referred to as the constraints on the choice set in the present that are derived from historical experiences. Understanding the process of change entails confronting the nature of path dependence in order to determine the nature of the limits of change that it imposes in various settings (North 1990). Japan's energy future is path dependent. It is embedded in a specific political, economic and social context, constrained by Japan's existing energy system, but also affected by changes in the global energy system. A movement away from the present pattern of energy use is constrained by a combination of three domestic sources of path dependency: institutions and organizations; interests; and beliefs and perceptions (or ideas). At the same time, the path of Japan's energy future is not quarantined from broader global energy trends. In fact, other countries face similar challenges to Japan and Japan's future choices are embedded in energy sector development outside Japan. The literature on energy transitions is essential in understanding the slow and protracted pace of global energy transitions, which also apply to Japan. Japan's future energy strategy and the pace of future energy transition are contingent on the interplay between these four sources of path dependency.

Institutions are commonly defined as the rules of the game, or the humanly devised constraints that structure human interaction. They are made up of formal constraints (such as rules, laws, constitutions), informal constraints (such as norms of behaviour, conventions, self-imposed codes of conduct) and their enforcement characteristics. Organizations comprise a group of individuals bound by some common purpose to achieve objectives. Organizations include political bodies (political parties, regulatory agencies), economic bodies (firms, trade unions), social bodies and educational bodies (North 1990). Japanese energy policy and its future direction are embedded in the country's institutional and organizational structure, with METI as the energy policy-making hub, the nuclear industry and the utility monopolies at the centre. This energy policy-making structure has remained remarkably stable for almost four decades (Moe 2012) and remains in place despite of the societal pressure to move away from nuclear power since the Fukushima disaster (Vivoda 2012). Against this backdrop, this chapter examines the main sources of path dependency, which affect and/or constrain Japan's future energy policy options. First, the chapter explains the institutional structure of Japan's political economy and energy policy-making process, which remains dominated by vested interests centred on METI. Second, the chapter discusses key decision-makers' interests and ideas, which continue to inform Japan's energy policy. Finally, the chapter engages the literature on energy transitions to highlight the path dependent nature of Japan's future energy use.

Japan's Political Economy, Vested Interests and Energy Policy Process

Hall and Soskice (2001) draw a distinction between two ideal types of political economies, liberal market economies and coordinated market economies. In coordinated market economies, firms depend more heavily on non-market relationships with other actors and to construct their core competencies. These non-market modes of coordination generally entail more extensive relational or incomplete contracting and more reliance on collaborative, as opposed to competitive, relationships to build the competencies of the firm. The equilibria on which firms coordinate in coordinated market economies are more often the result of strategic interaction among firms and other actors than the outcomes of demand and supply conditions (Hall and Soskice 2001: 8).

Japan is a distinctive 'variant of capitalism' with a strong institutional frame and is, according to Vogel (2006: 8), a coordinated market economy, as it fosters long-term cooperative relationships between firms and labour, firms and banks and between different firms, to produce relatively stable networks of business relationships (*keiretsu*). In this system of governance, the bureaucracy plays a critical role in protecting industry from international competition, promoting industry through an active industrial policy, managing competition in sectoral markets and establishing and maintaining the framework for private sector coordination. Industry associations have historically served as important intermediaries between the government and industry (Johnson 1982, Samuels 1987). In the Japanese system, tight coordination between government and business has been accepted as the natural order. Regulation of industry has been strategic; it was not considered distinct from the promotion of industry (Yergin and Stanislaw 2002: 145). The incentives for regulators to effectively implement regulations can become distorted when industry actors exert undue influence over the regulatory process.

Understanding Japan's political and economic model of governance would be incomplete if its traditional practices are excluded from the analysis. With *amakudari* (descend from heaven) and *amaagari* (ascent to heaven), a revolving door between many branches of the Japanese government and corporations has become a widespread practice. *Amakudari* allows government bureaucrats to take up lucrative positions at the companies they once oversaw when retired. On the other hand, *amaagari*, a less known practice, allows Japan's government agencies to freely hire experts from industrial sectors (Schaede 1995, Horiuchi and Shimizu 2001). In an analysis of *amakudari*, Colignon and Usui (2003) hold that one of the reasons *amakudari* survives is because the central bureaucracy needs it. More concretely, as the central bureaucracy is small, many projects are outsourced to the local governments, which create sectionalism among the central ministries. Importantly, the central bureaucracy emphasizes the use of industry associations and public corporations for specific projects instead of expanding existing ministries. No industry has historically been as rife with *amakudari* and *amaagari* as the nuclear power sector in Japan (Wang and Chen 2012) and this will be explored in more detail in Chapter 6.

Scholars have provided numerous interpretations on the vested interest theme. The notion that industries seek to utilize the state regulations and institutions for their own benefit, rather than for the economy as a whole, can be found in Stigler's (1971) classic article on regulatory capture. Industries use their political influence to control entry into their own industry, as well as control the rise of industries producing substitutes for their own goods. Others have focused on the institutional fit of the system. Institutional theory tells us that institutions create stability. They are the rules of the game, leading to path dependencies, acting as bulwarks against radical change (March and Olsen 1989, North 1990, Olson 1982). But consequently, institutional change tends to happen at a far more glacial pace than technological change, a point described as early as in Ayres' (1944) work on institutional lags.

It is very easy for a society to be locked into economic practices and institutions congruent with the needs and requirement of old and established industrial actors (Gilpin 1996). In energy, this problem has been widespread for centuries, particularly because the sector contains some of the world's biggest (and politically most influential) industrial giants. Unruh (2000) talks about techno-institutional complexes – large technological systems embedded through feedback loops between technological infrastructure and institutions. Once locked in, they are exceedingly hard to replace, because so many powerful interests have invested in the perpetuation of the system.

While combining Schumpeter's (1942, 1983) evolutionary economics with Olson's (1982) notion of vested interests silting up rigidities in the economy, Moe (2012) proposed a theory of structural economic change. Moe accounts for how industries seek to influence the political authorities and how institutional structures, once established, solidify, held in place by industrial, political and institutional interests, making radical change to the system ever more difficult. According to Moe (2012), structural change routinely meets with resistance from vested interests that feel threatened.

A successful energy regime produces vested interests seeking the perpetuation of that regime. The stronger the interests, the harder it becomes for the state to pursue structural change. Consequently, vested interest structures in the energy sector need to be analysed in order to determine why different sectors, actors and industries are being treated so differently by the Japanese state. While vested interests are not the sole relevant variable, they are at the heart of understanding Japanese energy policy (Moe 2012). Vested interest structures consist of more than just concrete interest groups seeking to influence concrete issues. However, with a specific policy-area and a limited time-period, as here, the main actors are more easily defined.

The Japanese government plays an active role in shaping the country's energy mix. It regularly outlines energy policy plans for the future with stated goals. The government's view of what energy security entails for Japan can be gauged from these reports. Such an active government approach to the nation's future energy structure is in stark contrast to the laissez faire approach of the US government, which largely leaves the transformation of the US energy demand mix to the markets. Japan produces a comprehensive national energy plan on a regular basis and has the most systematic and comprehensive energy planning process, not rivalled by any of the other major advanced industrialized countries. The Japanese policy process is based upon a slow, mid-level-bureaucratic, group-consensus process that emphasizes continuity and the priority of maximizing Japanese economic interests. The Japanese policymakers apply a passive, adaptive process to new situations; a tactic that is pragmatic and not overly reactive. A close but informal consultative mechanism between industry and government that seeks to maximize market forces is also involved.

Government agencies are the main actors in energy policy. Continuity in government, especially in the bureaucracy, helps explain why Japan has followed a fairly consistent trajectory in its energy policy since the oil crises. The policy planners have been the same people over the years. By the time a policy paper reaches the cabinet-ministerial level energy advisory councils and other decision-makers, a government- and industry-wide deliberative and consultative process has already taken place that virtually ensures the recommendation of the policy proposals.

The key administrative organization responsible for this process is the METI, which has central responsibility for most areas of energy policy. The 1973 oil crisis led to an important shift in the relative power of those participating in shaping the energy future of Japan. In particular, the Ministry of International Trade and Industry (MITI) emerged as a central coordinating player in the energy game. MITI was a major proponent in establishing the Agency for Natural Resources and Energy (ANRE) in 1974 in order to expand the bureaucratic infrastructure necessary to achieve policy objectives. Prior to 1974, energy policy was formulated and managed by the Mines and Coal Bureau and the Public Utilities Bureau in MITI. In order to prevent the return to pre-1974 situation, in 1977, MITI preserved its bureaucratic position by successfully resisting Prime Minister Fukuda's efforts to create a separate energy department (Nye 1981: 212). In fact, MITI was careful to incorporate the ANRE into its own organizational structure in order to exert considerable influence over the agency and the energy policy-making process (Lesbirel 1988: 288). The ANRE remains the focus of Japan's energy policies.

MITI/METI is known for actively seeking advice from large business entities and even requiring such an interaction on a near-constant basis (Yergin and Stanislaw 2002: 146). In changing from MITI to METI this core institution of the Japanese system has not abandoned its tradition of providing support to Japanese industry and working closely with firms (Vogel 2006). After reorganization in 2001, the new METI continued to implement energy policy on nuclear, fossil and renewable sources. With its historically very strong links between industry and bureaucracy, the Japanese bureaucracy often has strong preferences and major policy-influence. Typically 'tribes' within METI actively represent their most important 'clients', that is, one particular branch of industry. Thus, the Japanese energy vested interest structure also comprises parts of the bureaucracy.

In the consultative process, the Ministry of Education (MoEd; university and research funding), the Ministry of Environment (MoE; pollution and greenhouse gas emission policy and regulation), the Ministry of Finance (MoF; budget support for national energy policies), the Ministry of Foreign Affairs (MoFA; diplomatic relations with energy suppliers) and several other bodies have a significant input. The MoF has the critical say on funding for energy R&D and budgetary support for energy programs and consequently has much leverage against METI. The MoE is the department with the stronger viewpoints, but also one which has little influence except for during special windows of opportunity. The Japan Atomic Energy Commission (JAEC) and since 2012, the Nuclear Regulation Authority (NRA), decide on matters related to research, development, utilization and safety of nuclear energy, including regulatory and licensing matters. The Japanese parliament has special committees on energy policy in both upper and lower houses.

The second major group is a loose alliance of business and industry leaders. Private industry has an input into national energy policy through its participation in government advisory bodies and through industrial federations or specific industrial lobbying groups, such as the Federation of Electric Power Companies (or the *Denjiren*). The *Denjiren* have been adamantly opposed to any rival power-generating actors and they own both nuclear and thermal facilities. The electric utilities are METI's main client and, consequently, there is a strong convergence of interest between the two. With personnel moving back and forth between the utilities (including nuclear) and METI, the utilities' interests are to a considerable extent entrenched in the bureaucracy. The electric utilities have traditionally been among Japan's most profitable and influential companies, reflecting their close relationship to the powerful industrial bureaucracy.

The electric utility regulator, METI, takes a decisively pro-business approach in their dealings with the electric power companies. In recent years, the MoE has competed with METI for the upper hand in the regulatory control of the sector (Peng Er 2010). While the MoE takes an actively pro-environment approach for obvious reasons of self-interest and preservation, the political necessity of maintaining a stable supply of power in Japan sometimes forces the MoE to turn a blind eye to certain environmental regulations in the name of economic efficiency and stability. This government–business relationship is strengthened via *amakudari*, which is an omnipresent phenomenon in the electric power sector; all major listed utilities have at least one former career bureaucrat sitting on the board of directors and elsewhere (Scalise 2012a: 149). Virtually all of Japan's regional business federations have electricity company CEOs at their heads (Calder 2012: 183).

Even though the Japanese government is responsible for the development of energy strategy, it is not an actual player in the market. This role is occupied by various semi private and private actors, which besides the electric utilities, include oil and gas companies, such as Inpex, JX Nippon, Osaka Gas or Tokyo Gas and trading houses. The large Japanese trading companies, such as Mitsubishi, Mitsui, Marubeni, Itochu or Sumitomo, handle much of Japan's energy imports. General trading companies, steel and banking also have strong interests in energy sector development. These companies are internationally distinctive in their ability to profit from diverse export, import and investment transactions and are especially important as energy development project catalysts (Calder 2012: 183). Their views are reflected through a wide range of industrial and corporate affiliations, including through the *Keidanren* (Japan Business Federation).

Third, the bureaucracy has had a very strong relationship with the Liberal Democratic Party of Japan (LDP) – the party apparatus, not the government. The LDP has traditionally consisted of multiple rival factions. It has been described as exchanging 'votes for money, money for favours, favours for positions, positions for patronage, then patronage for votes, and so on' (Castells 2000: 232). 'Almost every interest-group imaginable is represented within its ranks, and it takes care to look after them' (Grimond 2002). Traditionally, the LDP is a strong supporter of nuclear power. The LDP has received sizeable donations from Japan's major nuclear plant makers – Toshiba, Hitachi and Mitsubishi Heavy Industries. The deep-pocketed regional utility monopolies and industrial energy users have cultivated salubrious ties with influential politicians through generous campaign contributions that far outpace the resources available to environmental groups (Duffield and Woodall 2011: 3748). Thus, while the main actors who determine energy policy are industry and bureaucracy, one cannot completely ignore the Japanese politicians and parties.

The Japanese vested interest structure is a genuine *iron triangle* of politics, bureaucracy and industry (Yergin and Stanislaw 2002: 145). Insiders are systematically locked in and protected, at the expense of outsiders (Katz 2003). As new issues or initiatives arise (e.g. restructuring the electricity market, overseas competition for energy, increased cost of energy imports, safety of nuclear reactors), these groups or organizations make adjustments, relocate resources, establish new committees and call on different experts to examine and report as appropriate. Through exclusive reporters' clubs (or *kisha kurabu*) these vested interests also have control over the direction of public debate or public opinion on energy-related issues through the media. The power of the vested interest structures will become evident in Chapter 6 (nuclear energy) and Chapter 7 (renewable energy), as they remain largely committed to nuclear power and opposed to a greater penetration of renewable energy.

Finally, there are separate organizations, experts, citizens groups and the media; all interested in energy issues, but generally located on the fringes of energy policy-making. Environmental and local citizen associations opposed to nuclear and other types of energy facilities have been fairly effective in delaying or stopping a number of projects. These groups have historically been Japan's major forces for change in energy policy. Although these groups enjoy limited political backing from major political parties, they still must be taken into consideration in the government planning process.

For the past four decades we have seen fairly remarkable stability in the structure of Japanese energy policy-making. The power of the utility companies and their close relationship with METI, in particular the ANRE, has remained constant. While long-term energy policy is based on input from experts on METI's Advisory Committee for Energy and various related deliberative councils, METI's bureaucrats design policy. METI still sits at the hub of energy policy-making, shrugging off any effort to wrest policy-making away from it (Moe 2012).

Interests and Ideas

In line with the broader functioning of Japan's political economy, evident in the vested interest structure between government bureaucrats, politicians and industry groups, Japan's energy policy-makers believe in government intervention in energy markets in order to enhance energy security. The Japanese government remains heavily involved in the functioning of Japan's energy markets and it remains closely linked with business groups, energy companies and trading houses. While there are some in Japan who hold dissenting views and argue for greater energy market liberalization, the evolution of Japan's energy policy since the oil crises indicates that those who favour government intervention remain in control of the policy-making process. With the exception of the liberalization of Japan's downstream oil industry in the mid-1990s (Horsnell 1997), no other pro-market government policy has been implemented since the oil crisis. While some piecemeal measures towards more liberalized electricity and gas markets were adopted during the 1990s, these markets remain heavily regulated and, in the case of electricity market, dominated by ten regional electric utility monopolies, which control both power generation and transmission. A partial gas market liberalization of the late-1990s resulted in the same electric utilities entering the market.

Japan's energy policy-makers view competition for energy in zero-sum terms, where one state's gain (of energy resources) is another's loss. Following this perspective, Japan faces particularly acute energy competition due to the country's poor resource endowment and geography. And being almost completely dependent on imported fossil fuels has meant that Japan has little margin for error in dealing with harmful actions by other players in the energy arena. Because of the energy market's inherent vulnerability, Japan's energy policy-makers have little confidence that its forces will balance supply and demand on a regional – let alone – global level. This is due in part to characteristics specific to energy economics: energy projects often require massive capital investment and long lead times. The market is also affected by the strategic importance of energy resources – which are both essential to industrial economies and the object of geopolitical tension and rivalry.

For these reasons, Japan's energy policy-makers see an important role for the government in shaping Japan's supply and demand patterns. While they recognize that government intervention may distort markets and constrain competition, they consider these concerns to be outweighed by the dangers of overreliance on markets. Pointing to boom-and-bust cycles in energy markets – where feasting (low prices and depletion) results in famine caused by inadequate investments due to low prices – they question the assumption that markets will efficiently allocate resources. Because of the heavy capital investment and long lead times associated with developing oil, gas, coal, or nuclear energy, cycles can emerge that cause dangerous mismatches in supply and demand.

This viewpoint holds that the private sector alone cannot overcome Japan's fundamental energy vulnerabilities; it must be supported and augmented by strategic state intervention. Likewise, the best fuel and technology mix for the nation is unlikely to be achieved through reliance on markets. Japan is better off if the government works hand-in-hand with industry to promote secure energy resources. Nuclear power holds a particularly important role in this view. Historically, Japan's energy policy-makers have been strong supporters of nuclear power. This includes conventional light water reactors (LWRs) and pressurized water reactors, which make up Japan's civilian nuclear fleet, as well as fast-breeder technology that can create more fissile material than it consumes. Japanese energy policy-makers are sceptical about the benefits of energy institutions and initiatives designed to promote regional cooperation. While they do not reject participation in international institutions, they tend to be sceptical about the ability of such institutions to make a meaningful contribution to Japan's energy future.

The oil crises of the 1970s marked a major shift towards government intervention in the energy markets. Reducing Japan's external dependency played an important role in the move to more active government management of Japan's energy supply and demand mix. The surge in oil prices over the past decade and the increasing competition with China for energy supplies has given further impetus to government intervention. A general view among Japanese energy policy-makers is that a 'leave it to the market' approach is not a solution since this could render oil and gas security more vulnerable (Nakatani 2004: 416). Instead, energy in Japan remains conceptualized as a national security issue (Phillips 2013: 25). The Fukushima disaster and the ensuing energy crisis have cemented this view. The government intervention in the energy market is seen as essential in order to ensure sufficient energy supplies. The vested interest structure centred on METI, the LDP and the electric utilities remains the main driver of Japan's energy strategy based on high level of government regulation of the energy market that favours status quo, in which semi-autonomous electric utilities and trading houses remain predominant market participants and gain diplomatic support in their overseas ventures.

Path Dependence, Energy Transitions and the Evolution of Japan's Energy Supply–demand Balance

The global energy system is in the early stages of a transition from carbon intensive fossil fuels to a variety of substitutes, bringing economic, strategic and environmental risks.

Scholarship on energy transitions suggests that these transitions have been both gradual and complex. As Grübler (1991) notes, 'Along its growth trajectory, an innovation interacts with existing techniques, and changes its technological, economic, and social characteristics. Decades are required for the diffusion of significant innovation, and even longer time spans are needed to develop infrastructures'. Smil (2008) makes the point even more concisely: Energy transitions 'are prolonged affairs that take decades to accomplish and the greater the scale of prevailing uses and conversions the longer the substitutions will take'. Coal had been in use for thousands of years, but it was not until growing urbanization led to a shortage of wood that the use of coal became more commonplace. Similarly, oil derivatives were used in lamps throughout the nineteenth century, decades before they became the world's dominant source of energy.

An examination of historical energy demand trends in Japan (Figure 1.1) and globally (Figure 1.2) reveals that substantial changes in proportions of energy use from various sources take decades. It was only as a consequence of discoveries of a superior source of energy that a relatively rapid transition to a new energy source has ensued, as in the cases of coal and oil. Occasional supply shocks, such as the 1970s oil crises, only marginally affect the historical pattern, with return to pre-shock shares within two decades. Currently, fossil fuels make up 87 per cent of global primary energy supply – constant for over two decades – with no serious competitors on the horizon.

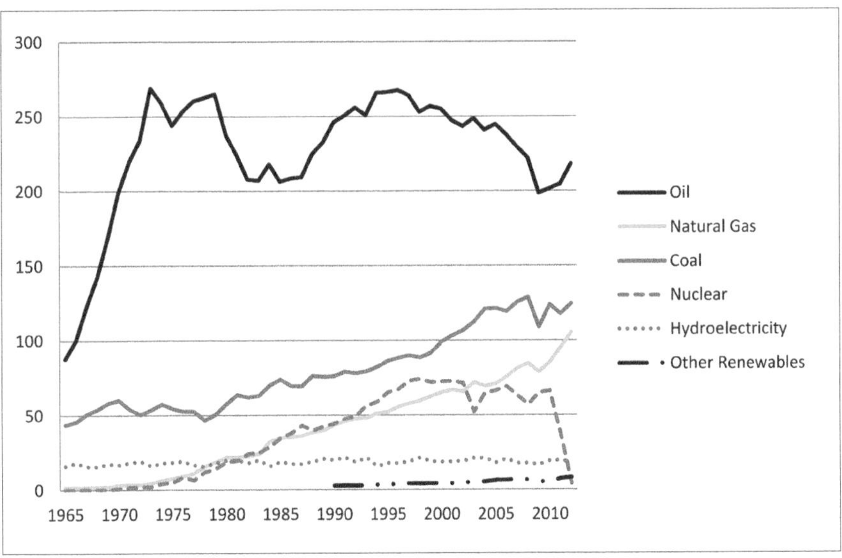

Figure 1.1 Japan's primary energy demand by fuel source (1965–2012; mtoe)
Note: 'Other Renewables' from 1990.
Source: BP 2013.

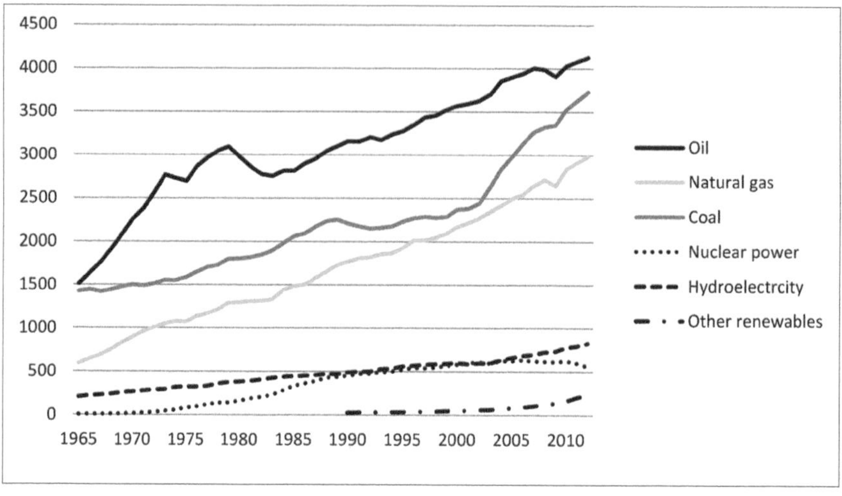

Figure 1.2 Global primary energy demand by fuel source (1965–2012; mtoe)
Note: 'Other Renewables' from 1990.
Source: BP 2013.

Historically, Japan's primary energy supply has been dominated by fossil fuels and specifically oil. In fact, pre-1973 oil crisis, oil accounted for 78 per cent of Japan's energy demand. Following the oil crises of the 1970s, the share of oil in Japan's energy demand continued to drop, reaching the minimum of 40 per cent in 2010, before rebounding in 2011 and 2012 as a consequence of increased demand for fuel oil due to the Fukushima disaster and the ensuing shutdown of most nuclear reactors. While coal provided approximately 30 per cent of Japan's primary energy supply in 1965, its share has dropped to as low as 13 per cent by 1978, mainly as a result of the uptake of nuclear energy. However, both its relative share and absolute volumes increased since and in the past decade coal supplied over 20 per cent of Japan's primary energy demand each year. The demand for natural gas shows a steady increase in both share of total energy demand and absolute volumes, reaching maximum values in 2012. Nuclear energy reached a maximum share of 15 per cent of Japan's overall primary energy demand in 1998, before dropping to a pre-Fukushima level of 13 per cent. The share of hydroelectricity stood at over 10 per cent in the mid-1960s. However, its share dropped significantly since and has remained at between 3 and 5 per cent since the early 1970s. The share of other renewables is very low, although, similar to natural gas, it has increased in both relative and absolute terms over the past two decades (Figure 1.3).

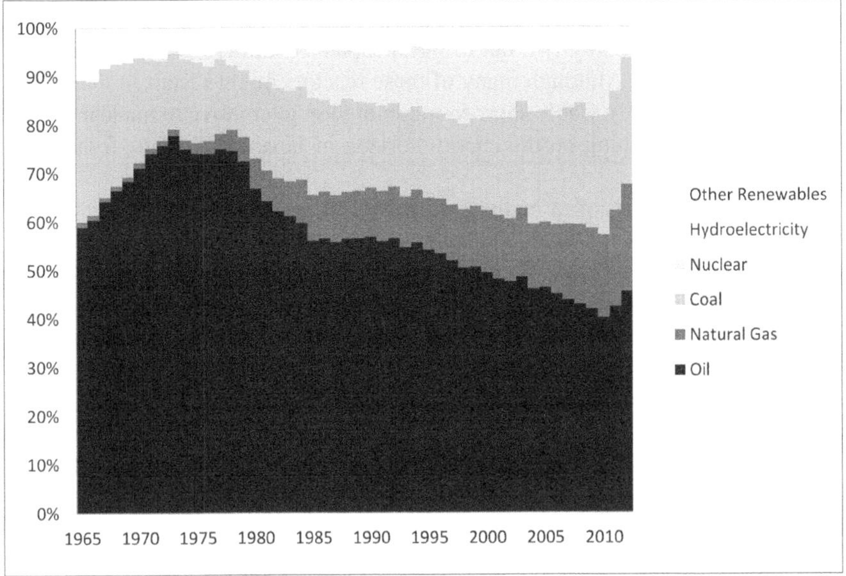

Figure 1.3 Japan's primary energy demand by fuel source (1965–2012; relative share)

Note: 'Other Renewables' from 1990.

Source: BP 2013.

The evolution of Japan's energy demand structure reveals that the government's measures to diversify away from oil in the aftermath of the two oil crises in the 1970s have been successful. Alternative energy sources such as coal, LNG and nuclear energy were emphasized (Scalise 2004: 161). These three energy sources largely made up for the reduction in Japan's oil demand during the 1980s and early 1990s. While by 1995 oil demand returned to its 1970s peaks, it dropped significantly since. Yet, regardless of government policy to reduce Japan's dependence, oil remains the dominant energy source in Japan with 45.6 per cent of the overall demand. Oil is followed by coal with 26 per cent and natural gas with 22 per cent. In aggregate, fossil fuels account for over 93 per cent of primary energy demand. The share of other energy sources is relatively low, with hydroelectricity in 2012 providing 4 per cent, other renewables under 2 per cent and nuclear power under 1 per cent of total supply (BP 2013). Japan's energy end-use remains dominated by its industrial sector, which takes up about 40 per cent of total current end-use energy consumption. The transportation sector accounts for 24 per cent of consumption, with the residential and commercial sectors accounting for the remainder of total energy use (IEA 2012).

The realities of energy transitions and the particularities of Japan's energy system hinder any quick move away from fossil fuels. Japan has reduced its nuclear power output and this reduction is likely to remain for the foreseeable future. In early 2013, only two out of 54 of Japan's commercial nuclear reactors have been operating. Although many of these reactors might restart in the future, fossil fuels are the only viable short-to-medium term alternative to nuclear power. There is, in fact, nothing on the energy horizon in Japan to displace fossil fuels (Smil 2010).

In addition, short of a major technological breakthrough, which makes renewable energy competitive with other energy sources on a large scale, it will take decades before renewable energy becomes competitive with fossil fuels in electricity generation and transportation sectors. A glance at past global energy consumption trends (Figure 1.2) indicates that, with the exception of hydroelectric power, renewable energy is a newcomer. Other renewable energy sources are negligible as sources of energy in the current global energy system. The same applies for Japan, where they start from a very small base (Figure 1.1). While the share of renewable energy in Japan's energy mix will grow, this is likely to materialize at a very slow pace due to relatively higher costs and other structural impediments (discussed in Chapter 7) that inhibit a fast uptake of renewables.

Conclusion

This chapter examines the three main sources of path dependency – interests, institutions and ideas – which affect and/or constrain Japan's future energy policy choices. It demonstrates that Japan's energy policy-making process has been dominated by the iron triangle of bureaucrats, industry groups and politicians. These vested interests, centred on METI, the *Keidanren*, *Denjiren* and the LDP, have remained in control of Japan's policy-making apparatus for at least four decades. The predominant view among these vested interests is that the Japanese government has to maintain an interventionist role in the functioning of the energy markets, as Japan – a resource poor country – is vulnerable to both external and internal energy shocks. This chapter also demonstrates that energy transitions are slow and protracted affairs and are another source of path dependency. More specifically, the slow nature of energy transitions serves as a major structural constraint to a rapid shift in Japan's energy mix in the future. Against this backdrop, the chapter briefly outlines the evolution of Japan's energy supply and demand balance, arguing that a quick move away from fossil fuels and towards renewable energy is unlikely.

Chapter 2
The Evolution of Energy Security and Energy Policy in Japan

Introduction

Historically, energy security has been most commonly defined as reliable supplies of energy at reasonable prices to support the economy and industry (Dorian et al. 2006). It has been perceived as a condition obtaining when national governments perceive that they will have reliable, affordable and uninterrupted access to the energy services necessary to maintain normal economic activity (Deese 1979: 140). In the past, most studies have conceptualized energy security in terms of security of oil supplies (Fried and Trezise 1993, Stringer 2008). This oil supply-based focus has as its cornerstones reducing vulnerability to foreign threats or pressure, preventing a supply crisis from occurring and minimizing the economic and military impact of a supply crisis once it has occurred. These goals implicitly assume an 'oil supply crisis' as the focus of energy security.

Yet, the last decade has seen an extraordinary shift in energy security challenges that challenge existing policy orthodoxies (Victor and Yueh 2010). Yergin (2006) noted that the traditional understanding of energy security is too limited and must be expanded to include many new factors and challenges, while at the same time recognizing that energy security does not stand by itself but is lodged in the larger relations among nations and how they interact with one another. The substance of these challenges therefore needs to be incorporated into a new concept of energy security. With increasingly global, diverse energy markets and increasingly transnational problems resulting from energy transformation and use, old energy security rationales are less salient and, other issues, including climate change and other environmental, safety, economic and international considerations are becoming increasingly important.

While the concept of energy security has many meanings, what is clear is the importance of secure sources of energy supply for our present way of life and the threats posed to that way of life when energy supplies are not secure. These threats are geopolitical, economic, technical, psychological and environmental. First and foremost, it is political factors that pose perhaps the greatest threat to energy security. In particular, as oil is unevenly distributed, political turmoil in the Middle East may affect oil supply. Second, economic factors can constitute a threat. Increases in oil and gas prices inevitably pose a threat to import-dependent consumer states such as Japan. Third, when we consider energy security, technical threats have to be overcome and safety at all stages of energy-related activities including development, transport and distribution, has to be secured. Safety of nuclear facilities has to be secured at all costs. Fourth, psychological factors cannot be ignored. Even if technical threats are overcome and safety is secured, this does not imply that anxiety among people about safety is removed. This apprehension remains as a psychological threat. Indeed, consumer panic may cause excessive demand and lead to a supply crisis, as was the case in Japan in 1973. Finally, environmental factors have to be considered in terms of energy security, as human activities cannot be accomplished without due concern for the environment.

Against the backdrop of a greater diversity of threats to energy security, the scholarly debate over the past decade has produced a more comprehensive and broader understanding of the concept of energy security. While acknowledging the polysemic nature of the concept (Chester 2010) and that energy security is primarily a matter of governments' subjective perceptions (Phillips 2013), the book adopts UNDP's (2004) definition of energy security as the availability of energy at all times in various forms, in sufficient quantities and at affordable prices, without unacceptable or irreversible impact on the economy and the environment. This definition will serve as basis for analysis of Japan's energy security challenges in the rest of the book. The post-Fukushima energy crisis in Japan all but confirms the validity of a broader conceptualization of energy security. Such broader understanding of the concept highlights the salience of factors other than security of supply. Energy challenges are multidimensional and a holistic approach is necessary to analyse the dynamics and how energy-related factors affect security, safety, economy and the environment. While the debate moved in this direction pre-Fukushima, the disaster simply reaffirms this broader understanding.

Consequently, this chapter examines the historical evolution of Japan's energy policy and energy security strategy. In surveying the evolution of Japan's energy policy, this chapter demonstrates how Japan's conceptualization of energy security evolved away from oil security, which is in line with the scholarly debate. The analysis is divided into four major periods: the post-war (1950–1973), the oil crises (1973–1979), diversification away from oil, energy efficiency and reducing greenhouse gas emissions (1980–2011) and post-Fukushima developments. Such a historical survey is essential to place current challenges in the context and to set the scene for subsequent chapters, which examine major challenges that pertain to various energy sources. The analysis of the evolution of Japan's energy policy in response to the 1970s oil crises is important as it speaks to Japan's past responses to exogenous shocks.

1950s–1973: The Price of Oil

From the mid-1950s it was evident that Japan's indigenous energy resources were inadequate to meet its objective of rapid economic growth, with its emphasis on heavy industry. The falling price of oil from 1958 reinforced the need to turn to imported oil as the basis for industrial expansion. The same applied to raw material imports; but it stood in sharp contrast to the tight government control maintained over imports of goods, capital and technology. An assumed elastic world supply of oil and rapid export expansion appeared to justify a largely passive non-interventionist policy with regard to energy. Such measures as were introduced were designed to cope with particular problems. A coordinated response to the risks of import-dependence seemed unnecessary, given the underlying assumptions about long-term fuel and raw material prices, political and economic interdependence with the US and the attraction of Japan's growing market to foreign suppliers of fuel and raw materials. The expectation of long-term oil abundance and the continuation of warm US–Japan relations implied that Japan's reliance upon crude oil supplied by the international companies appeared to involve little risk. No foreign oil supplier was likely to jeopardize its share of the world's largest and fastest-growing import market (Surrey 1974).

During the 1960s and prior to the oil crises Japan moved away from autarkic policies designed to bolster its domestic coal industry to policies that lead inexorably to its almost total dependence on imported energy. Such a move was sparked by the insight of Japanese policy-makers that using imported oil to fuel Japan's economy would ultimately bring production costs down, thereby stimulating exports (Hein 1990). This energy situation contributed substantially to the high growth of Japanese industries (especially energy intensive industries such as iron, steel and petrochemicals) that was achieved in the 1960s. These industries became highly competitive in the international market (Oshima et al. 1982: 87).

During the 1960s, as Japan experienced rapid economic growth, the government's primary policy goal was to secure a cheap supply of oil (Fujii 2000: 60). This period was characterized by government encouragement of the local oil refining industry, which was heavily protected and regulated, under the *Petroleum Industry Law* (1962). It also promoted the rapid growth of energy intensive and export orientated, heavy industry (chemicals, iron and steel, aluminium, etc.). The increasing price competitiveness of imported oil compared to local coal, led MITI to reduce its support for the domestic coal industry. Nevertheless, the government enforced a system of protection for the coal industry, which enabled it to sell its coal well above the international price and required major coal consumers to purchase a designated proportion of their coal demand from local mines. This was the period of Japan's most rapid post-war growth.

Before 1973, Japanese energy policy concentrated primarily on price, not security of supply. Low energy prices were essential to government's strategy of encouraging investment in heavy, energy-intensive industries (Tsurumi 1978). Cheap imported oil was favoured over indigenous coal or hydroelectric resources. As a result, Japan's dependence on energy imports rose from 26 per cent in 1955 to 46 in 1960 and 90 per cent in 1973 when the oil crisis struck (Nye 1981: 213).

1973–1979: Responses to Oil Crises

The oil crises of the 1970s had a significant impact on Japanese energy policy. Consequently, Japan's national energy policy took a dramatic turn with the 1973 oil crisis (Morse 1981: 1). As a country with a high reliance on imported oil and a high sensitivity to security matters, Japan commenced a long-term program of using substitutes for oil such as nuclear, coal, LNG and new energy technologies aimed at reducing the dependence on imported oil and protecting Japan against further increases in oil prices and unexpected interruptions in oil supplies. The development of alternative energy can be interpreted as an insurance policy to cover the risks associated with an over-dependence on Middle East oil (Lesbirel 1988: 287). This program was market-conforming but allowed for a significant role for government in longer term planning processes.

The fourfold increase in crude oil import prices between June 1973 and June 1974 forced an urgent re-evaluation of the priorities of energy policy and the appropriateness of Japan's industrial structure. In 1973 approximately 78 per cent of all Japan's energy needs were met by imported oil and, of this, 80 per cent came from the Middle East and a further 17 per cent of energy needs were fulfilled by imported coal, gas or nuclear fuels (BP 2013). Furthermore, because of the pattern of industrial growth in the 1960s and early 1970s, Japan's industrial structure was heavily skewed towards energy intensive basic industries. These were directly or indirectly export orientated. Hence the rapid rise in oil prices threatened to severely undermine the viability of Japan's industrial sector and external trade position. It was suggested that no nation was more profoundly affected by the 1973 oil crisis than Japan (Nemetz and Vertinsky 1984: 70). For example, the 1973 oil crisis caused a 7 per cent decline in GDP in Japan, compared to 4.7 per cent in the United States and 2.5 per cent in Europe (Salameh 2001: 131).

The oil crisis in 1973 had a serious impact on the Japanese economy and society. A rapid increase in commodity prices caused substantial economic confusion and a panic-like atmosphere prevailed. Naturally, energy became a major concern to the public and the government and energy policy in response to high prices and the uncertainty of oil supply was taken more seriously than in some other industrialized countries (Oshima et al. 1982: 87–88). The short-lived interruption in the flow of oil had an enduring influence upon the perception by the Japanese of their energy vulnerability. This experience led to a potential overestimation of the risks associated with disruption in the supply of energy and a continued sensitivity to events which could create threats to oil-supply security (for example, the Iran–Iraq War) (Lesbirel 1988: 288).

Amid the oil crisis, Japan took three important policy steps. First, it passed the *Petroleum Supply and Demand Optimization Law* (late 1973), which set oil-supply targets and restricted oil use. Reinforced later by supplemental tax measures, this legislation encouraged industrial consumers in energy-intensive sectors such as steel, paper and petrochemicals to economize on oil consumption, through such techniques as cogeneration. Second, it passed three laws in support of the nuclear industry (June 1974). These encouraged the rapid expansion of reactor use from three in 1970 (two experimental) to 46 in 1993. Third, it passed the *Petroleum Stockpiling Law* (1975), which provided financial assistance to private firms in maintaining a 70 day supply of petroleum products. Based on this law, on one hand, the government supplied financial support to industry for its own stockpiling. On the other hand, Japan National Oil Corporation (JNOC) became responsible for national stockpiling by the government (Morse 1981a: 46, Oshima et al. 1982: 102). Together, these three crisis-driven measures were the crucial policy steps generating the impressive diversification outcomes (Calder 2012: 185). Laws enacted between 1973 and 1975 enabled the government to control energy prices and supply.

These policies were implemented by a range of measures including interest rate subsidies, cash grants, legal regulations, administrative direction and indicative, or rather consensus, planning via MITI's long-term energy supply and demand outlooks. The latter outlooks were formulated after exhaustive consensus negotiations between MITI and energy users and producers. They became a powerful policy tool in encouraging energy conservation and fuel switching in the 1970s and early 1980s.

Moreover, the basic thrust of energy policy was altered in a policy document released in 1975 by the Ministerial Council on General Energy Policy headed by the Prime Minister. The Ministerial Council published a document entitled *Basic Direction of General Energy Policy*, which articulated the principles that have guided Japanese decision-making and set the basic direction of energy policy until 1986. The major components of this policy included the continued development of nuclear power, the promotion and revival of indigenous energy sources such as coal, the encouragement of energy research and development, increasing the efficiency of secondary energy utilization, diversification of overseas energy sources and the continued participation of Japan in foreign resource development and international cooperation (Nemetz et al. 1984/1985: 560). Security of supply was to be achieved by reducing dependence on oil and substituting non-oil energy sources; securing a stable oil supply; pursuing energy conservation; and undertaking research and development into new energy sources. In the long-term, reduced dependence on imported oil, particularly Middle East oil, was believed to be the best means of improving the security of energy supplies (Perkins 1994: 596). This indicated that in future, top priority would be given to security of energy supplies, to ensure the continued viability of the national economy. Previously this objective had been balanced by the desire to obtain energy at the most competitive prices.

The significantly higher prices of energy led to important readjustments in the use and composition of fuels consumed in Japan. Strong incentives for conservation and diversification were created by concentration of demand in a few industries, the high proportion of energy costs in their production processes and the expected costs of further supply interruptions. Conservation efforts resulted in a reduction of approximately 9 per cent in the energy input per unit of industrial product within five years (Nemetz et al. 1984/1985: 558). While much of the conservation that took place during the period between the two energy crises can be attributed to market forces in a highly competitive economy, the psychological impact of the first crisis and the shift it caused in the relative power of those with a stake in energy policy should not be discounted. The recognition of a collective threat brought about demands for coordinated action to secure Japan's economic position and continued development.

A vigorous energy policy regime was therefore in place prior to the 1979 oil crisis. The main objectives of these policies were to encourage energy (particularly oil) conservation by promoting an increase in the efficiency of energy use and a reduction in the energy intensity of economic activity and a switching from oil to alternative fuels and energy sources. In this way it was hoped that energy and, particularly oil, demand could be cut without sacrificing long-term economic growth.

The second oil crisis of 1979 was touched off by the interruption of petroleum supplies resulting from the Iranian revolution, initiated by the Organization of Oil Exporting Countries (OPEC) at its Geneva meetings of June 1979. The doubling of crude oil import prices between March and December 1979 and their steady drift up to peak by November 1982 resulted in a renewed crisis in the Japanese economy. At that time, Japan was in the midst of an intense process of energy conservation and substitution. This second crisis further reinforced the energy policies of the government and the fuel substitution and conservation efforts of the private sector. By 1980, energy security had become a major issue in Japanese politics (Nye 1981: 213).

The second oil crisis also had a substantial impact on Japanese industry and the economy, but the response of the public was much calmer because of the previous experience. In fact, on this occasion, the Japanese economy responded much better than in some other countries with respect to problems such as inflation, unemployment and foreign exchange, which were becoming serious problems in industrialized countries. Japan had been more flexible in adjusting to energy constraints than had been its industrial competitors, in part through greater gains in labour productivity and overall economic growth (Katzenstein and Okawara 1993: 107). In fact, Japan has managed to navigate through the second oil crisis better than in 1973. During the second oil crisis, wage increases were kept in line with low inflation rates, the national economy and the budget deficit were managed well and the necessary adjustments were made for higher oil prices (Morse 1981b: 3). This was in no small part due to the effective implementation of energy policy: reduction in energy consumption per GDP and fuel conversion from oil to alternatives has happened much faster than predicted (Oshima et al. 1982: 88).

During the second oil crisis, the major supply cuts were to the Japanese oil companies not directly affiliated with, but dependent upon, supplies from the major international oil companies. The companies suffering cuts scrambled to the spot market to make up for the shortfalls – this kind of disruption competition helped drive up the spot-market price for oil and strained relations among consuming nations.[1] The Japanese government responded to these cutbacks in three ways. First, MITI encouraged domestic oil companies and trading firms to secure crude oil directly from the oil-producing states. Second, MITI raised the price ceiling for spot-market purchases, eventually doubling Japan's spot-market purchases (Morse 1982: 261–62). Third, the increased sense of Japan's vulnerability following the second oil crisis triggered a heightened and relentless search for alternate sources of energy and a diversification of purchases internationally. The heads of two of Japan's most critical ministries, the Foreign Ministry and the MITI, spent much of 1979 engaged in 'fuel diplomacy' (Klein 1980: 43), or diplomatic activity designed to enhance Japan's access to energy resources and its energy security, which will be explored in more detail in Chapter 3.

Unlike the situation after the 1973 oil price rises, in 1979 oil lost its competitiveness compared to fuels like coal and LNG in electricity generation and a range of industrial users. The policy response was a considerable tightening of existing energy policies. The second oil crisis reinforced the need for accelerating Japan's energy diversification program. It intensified expectations that the development of alternative energy was not proceeding quickly enough. Government plans were articulated in the *Long-range Energy Supply and Demand Forecasts (Preliminary)*, published in August 1979. They called for a 12 to 17 per cent reduction in energy utilization through conservation practices, as well as an accelerated program of fuel substitution. In particular, emphasis was placed on the development of coal and LNG supplies, as well as nuclear and other non-conventional sources (Nemetz et al. 1984/1985: 567). In 1979 and 1980 laws were introduced to promote energy conservation and fuel switching (the *Law Concerning the Rational Use of Energy* and the *Alternative Energy Law*). The *Alternative Energy Law* provided MITI the legal means with which to implement diversification policy. The Law gave MITI more supervisory powers in energy markets (Lesbirel 1988: 290). In addition, these laws included a range of prescriptive and persuasive measures, as well as administrative mechanisms to achieve these objectives. For the industrial sector, the government provided tax benefits for investment in energy saving equipment and, for the residential and commercial sector, financial help for the installation of solar systems and for energy conservation measures in housing (Oshima et al. 1982: 101). At the same

1 The term 'spot markets' is used to describe transactions which involve the near–term purchase and sale of a commodity, such as crude oil and refined products. In the crude oil market, 'spot' contracts typically involve delivery of crude over the coming month, e.g., a contract signed in June for delivery in July. Spot markets are often referred to as the 'physical market' since they entail the buying and selling of physical volumes. These markets consist of many buyers and sellers, including refiners, traders, producers, and transporters, transacting throughout the chain of supply – from the oil well right through to the refinery. These markets provide the benefit of allowing buyers and sellers, e.g., refiners and marketers, to more easily adjust their crude supplies to reflect near-term supply and demand conditions in both the product markets and the crude oil markets.

time, several industry restructuring plans were implemented to shift capital and labour out of industries like aluminium smelting, which had lost their international competitiveness because of rising energy prices (Perkins 1994: 597).

1980–2011: Diversification Away from Oil, Energy Efficiency and Reducing Greenhouse Gas Emissions

The oil shocks of 1973 and 1979 brought down the curtain on Japan's post-war economic boom, which was characterized by double-digit annual rates of growth (Moriguchi 1988: 307). At the same time, Japan has been relatively successful in achieving its energy objectives and has been able to increase the share of alternative energy in primary supply. In response to the oil crises of the 1970s, Japan succeeded in implementing its plans to a degree that other governments found difficult to achieve (Vernon 1983: 97). As a result, Japan came out of the 1980s with a high degree of energy security for a country so poorly endowed with domestic resources (Stewart 2009: 181).

The major objective of the aforementioned August 1979 energy supply and demand forecasts published by the Japanese government was to limit Japan's oil imports to 6.3 million bpd by 1985 and to reduce oil dependence to 50 per cent by 1990. When published, the forecast was considered to be rather optimistic and the government described these figures as a working target, although they were approved by the Cabinet (Oshima et al. 1982: 99). Considerable success has been achieved in reducing oil imports and securing a switch from oil to alternative fuels. Indeed, Japan's oil imports in 1985 stood at 4.4 million bpd, a 22 per cent drop from 1979. Meanwhile, Japan's oil dependence dropped from 78 per cent in 1973 to 73 per cent in 1979 and further to 57 per cent in 1990. The reduction in oil's share has mainly been taken up by nuclear power and natural gas. The share of nuclear power in total energy consumption increased from only 0.6 per cent in 1973 to 11 per cent by 1990, while that of natural gas increased from 1.3 per cent to 10 per cent over the same period (BP 2013).

In response to the oil crises, the Japanese government adopted energy conservation and efficiency measures, which were among the most stringent in the industrialized world. Energy efficiency was a strategy initiated by MITI in 1983 (Fujii 2000: 60), which achieved much success in the following two decades. Despite the dramatic fall in oil prices, starting in 1986, the Japanese government remained committed to energy conservation and fuel switching. Both the government and the business sector anticipated medium- to long-term supply constraints and volatility in the oil market. This expectation was borne out by the continued instability in the Middle East, culminating in the Gulf War. Hence, all the policies developed after the 1979 oil crisis to promote fuel switching and energy conservation and efficiency remained in place at the beginning of the 1990s.

As a consequence of Japan's strategies to enhance energy efficiency, Japan now has one of the lowest rates of energy usage per unit of GDP produced among the industrialized countries. In the mid-1960s, Japan was referred to as a 'polluter's paradise' (Broadbent 2002), with air pollution, mercury poisoning and chemical pollution at alarming levels. Arguably, Japan's most notable energy policy success since the oil crises was in the area of energy efficiency (Calder 2012: 185), leading some to dub Japan a 'superpower' in energy efficiency (Stewart and Wilczewski 2009). The ratio of total energy requirements to real GDP was virtually the same in 1973 as in 1960, but fell by 27 per cent by 1982 (Bobrow and Kudrle 1987: 560). The 1977–1987 GDP increase of 42 per cent was accompanied in energy demand rising by only 14 per cent and energy intensity decreasing by 21 per cent (Luta 2010). As evident in Figure 2.1, Japan's energy efficiency was enhanced by over 30 per cent since the 1973 oil crisis (Masaki 2006, IEA 2008, Smil 2008, Stewart 2009, *The Economist* 2011, Valentine 2011). In fact, the Japanese industry currently uses a similar amount of energy as it did during the oil shock of 1973. From the 1990s, Japan has also attained the highest level of efficiency in thermal power generation, a level it still maintains (Sano 2011).

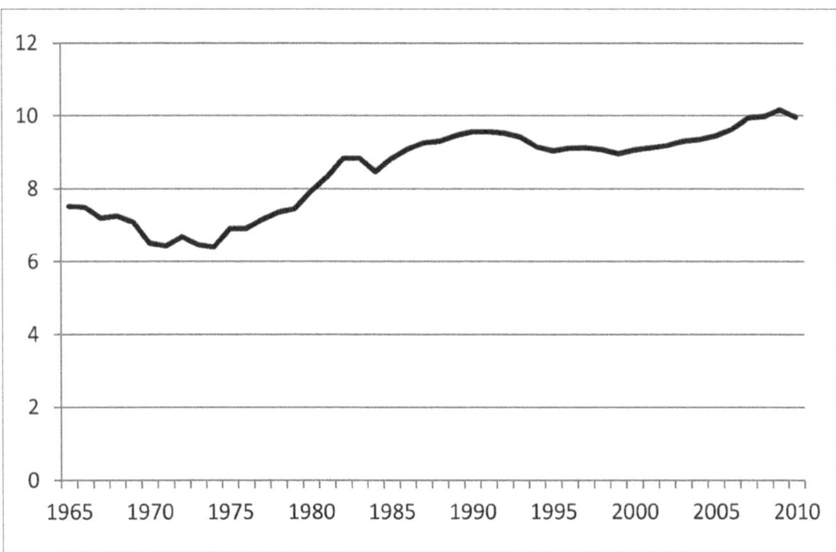

Figure 2.1 Japan's GDP (US$) relative to energy demand (kg of oil equivalent)
Note: GDP data are in constant 2000 US dollars. Dollar figures for GDP are converted from yen using 2000 official exchange rates.
Source: BP 2013, World Bank 2013.

The 1986 decline in international oil prices created market and political pressure for change in Japan's energy diversification policy. It exacerbated existing inter-fuel competition in energy markets between oil and non-oil sectors and intensified conflict over short-term economizing and long-term energy security objectives. The downstream oil sector and major industrial users of oil exerted considerable pressure on MITI to change energy policy and allow market forces to dictate policy outcomes. They argued that MITI should relax oil import restrictions and encourage the greater use of oil to reflect market prices and proposed that the taxation structure should be modified so as to reduce and, ultimately, abandon taxes which drive a wedge between marginal costs and prices. In essence, they argued that MITI should refrain from intervening in energy markets through the use of regulatory, budgetary and indicative planning mechanisms and allow competitive market forces to dictate the structure of energy supply and energy policy outcomes. MITI and the major alternative energy producers opposed policy modification for fear that a diversion from current policy would impede longer term objectives of reducing dependence on imported oil. These interests saw unacceptable economic and political costs associated with a major shift back into the consumption of oil in the Japanese economy. They opposed any change in energy policy in response to perceived short-term market fluctuations for fear that it would disrupt medium and longer term planning processes and adversely affect their bureaucratic resources and positions in the energy-policy-making process (Lesbirel 1988: 285).

Likely as a result of low oil prices and a well-supplied international market, during the late 1980s and 1990s, the public interest in Japan's energy security issues was low. In terms of importance of policy priorities, energy and environmental issues were ranked lower than issues concerning the ageing population and social security (Sudo 2008: 148). Deregulation of the Japanese economy began in the mid-1980s with the introduction of the *Maekawa Report*. With the end of the Cold War and advent of the economically stagnant 1990s, Japan's energy policy has moved away from one of pure 'dependency' and 'economic aid in exchange for oil' within highly regulated energy markets to a policy of gradual deregulation, market efficiency and alignment with the United States to obtain regional security in the maintenance of its oil supply flow (Scalise 2004: 159–60). In the 1990s, deregulation extended into the energy industry; first materially affecting the petroleum industry and later extending into the electricity and gas industries. For many years, Japan's energy industry was a textbook example of a regulated industry, but the picture has changed somewhat since the 1990s. A string of events has taken place, altering the operational landscape, including partial liberalization of the electricity and gas retail markets and the lapse of the *Refined Petroleum Import Law* (Scalise 2004: 162–63). Nevertheless, these reforms have not resulted in lower electricity prices and have not challenged regional dominance of electric utilities. In the late 1990s, Japanese government energy policy objectives were to: (1) improve energy utilization efficiency; (2) restructure the national energy mix to incorporate renewable energy sources; and (3) positively promote and pursue international cooperation in the energy field (IEA 2008).

The Kyoto Accord was signed in 1997 and the world's leading nations agreed to an historic agreement to reduce carbon and protect the earth's climate. Japan became a leader in this effort, especially through energy efficiency, promotion of nuclear power as a low emissions energy source and achievement of high thermal efficiencies in power plants. Consequently, Japan adopted another round of measures in the late 1990s and early 2000s, largely in response to growing concerns about climate change (Duffield and Woodall 2011: 3741). For example, in 2005, Japan introduced a voluntary emissions trading scheme that covers CO_2 emissions from fuel consumption, electricity and heat, waste management and industrial process from over 300 companies (Egenhofer 2012: 364).

More broadly, in June 2002, the Japanese government created a more systematic and comprehensive energy policy planning structure. In June that year, the Diet adopted a *Fundamental Law on Energy Policy Measures*, also known as the *Basic Act on Energy Policy* (Law No. 71), which set the general guiding direction for Japan's future energy policy (IEA 2008: 29). The Basic Act established three general principles of Japan's energy policy (the 3Es): securing a stable supply of energy, ensuring environmental sustainability and utilizing market mechanisms upon due consideration of the first two goals. While the Basic Act provided no specifics about energy policy, it required the government to formulate a basic plan to promote energy supply and demand measures on a long-term, comprehensive and systematic basis in line with the 3E principles (Japan Association of Petroleum 2012: 13). The government is tasked with reviewing the basic energy plan at least every three years and revising it as necessary in light of changing circumstances and the effectiveness of existing policies. METI was tasked with formulating the draft basic energy plan and then seeking cabinet approval before reporting it to the Diet (Duffield and Woodall 2011: 3742).

The first Basic Energy Plan (BEP) was adopted in October 2003. Its key points were to promote nuclear power generation, enhance efforts to secure a stable oil supply and lead the formulation of an effective international framework for enhancing energy conservation and coping with climate change (IEA 2008: 29). A revision of the BEP was prepared in late 2006 and adopted by the cabinet in early 2007, but it was based on and largely overshadowed by another energy policy statement, the New National Energy Strategy (NNES), which was issued by METI in May 2006 (IEA 2008: 59). The second BEP was mainly overshadowed by the NNES as it directly addressed the emerging energy competition with China in the context of rising oil prices, something that was overlooked in the BEP.

In contrast to the BEP, the NNES did not receive broader government approval, but it figured prominently in subsequent energy policy discussions. The NNES was developed in response to renewed concerns about Japan's energy security due, in particular, to rising oil prices, a revival of resource nationalism among foreign oil and gas suppliers and growing regional competition over energy resources, particularly with China (Atsumi 2007: 29, Duffield and Woodall 2011: 3742). Thus, in contrast to the Basic Act, the NNES placed primary emphasis on and sought to bring greater attention to the issue of energy security. During this time, the policy-planners and decision-makers were moving towards general agreement that Japan should also enter into the scramble for oil, mainly due to the fear of China (Atsumi 2007: 30).

To promote Japan's energy security, the NNES established five ambitious numerical targets to be attained by 2030 (METI 2006a: 14):

- a further 30 per cent improvement in energy efficiency (over 2003);
- a reduction in Japan's oil dependence from nearly 50 per cent to less than 40 per cent of the total energy mix;
- a reduction in the oil dependence of the transportation sector from nearly 100 per cent to around 80 per cent;
- a preservation or increase in nuclear power's share of electricity generation to 30–40 per cent or more; and
- an increase in the amount of oil produced by Japanese energy companies from 15 per cent to around 40 per cent of total oil consumption.

During the time of the drafting of NNES, a vigorous debate pitted those who see the country's energy security interests best secured through market mechanisms against those who favour strategic government intervention and championing – to the extent possible – energy autonomy. The NNES reflected a shift towards policies favoured by the autonomists, outlined in Chapter 1. This strategy not only sought to reduce Japan's external dependencies, but also favoured more active government intervention in shaping internal and external markets (Evans 2006).

During the following years, however, more of a balance was restored in Japanese energy policy. Concerns about security of supply abated somewhat, despite a continued rise in oil prices, while concerns about environmental sustainability, especially climate change, returned to the fore. In May 2007, Prime Minister Shinzo Abe announced an initiative, *Cool Earth 50*, to reduce greenhouse gas emissions in Japan and globally in the short-, medium- and long-term. In July 2008, the cabinet adopted a detailed *Action Plan for Achieving a Low-carbon Society*. Shortly after taking power in September 2009, the new government led by the Democratic Party of Japan (DPJ) announced an ambitious goal of reducing greenhouse gas emissions by 25 per cent below the 1990 level by 2020 and then prepared a detailed bill on Global Warming Countermeasures, which it submitted to the Diet the following March (Duffield and Woodall 2011: 3742).

In June 2010, the Japanese cabinet adopted a new BEP. This was the third such plan that the government approved since the passage of the *Basic Act on Energy Policy* in 2002 and it represented the most significant statement of Japanese energy policy since the publication of the NNES in 2006. The 2010 BEP was the last pre-Fukushima document that outlined the long-term strategy for the country's future energy mix. Perhaps more than its predecessors, the 2010 BEP established a number of ambitious targets as well as more detailed measures for achieving those targets. The targets included a doubling of Japan's 'energy independence ratio', a doubling of the percentage of electricity generated by renewable sources and nuclear power and a 30 per cent reduction in energy-related CO_2 emissions, all by 2030. This plan targeted the nuclear share of power production to surge from approximately 30 per cent to 50 per cent by 2030 (METI 2010a, Duffield and Woodall 2011).

The BEP established five ambitious targets for 2030:

- doubling Japan's 'energy self-sufficiency ratio (ESF)' (from 18 per cent) to about 40 per cent, and 'self-developed fossil fuel supply (SFFS) ratio' (from 26 per cent) to about 50 per cent, and consequently raising 'energy independence (EI) ratio' (from 38 per cent) to about 70 per cent;[2]
- raising the 'zero-emission power supply ratio' from 34 to 70 per cent;
- halving the CO_2 emissions of the residential sector;
- maintaining and enhancing the energy efficiency of the industrial sector; and
- maintaining or obtaining 'top-class' shares of global markets for energy-related products and systems.

The bulk of the 2010 BEP identified and proposed a number of specific measures for achieving these targets. The majority of the measures fall into three broad categories. Following on from concerns raised in the 2006 NNES, the first included measures to secure energy resources and to enhance the stability of supply. The second category consisted of measures to create an independent and environmentally friendly energy supply structure. This category included measures to expand the introduction of renewable energy sources, to promote nuclear power generation and to achieve advanced utilization of fossil fuels, especially coal. The third category consisted of measures for 'realizing a low carbon energy demand structure' (METI 2010a).

In terms of the basic goals it established, the 2010 BEP offered considerable continuity with previous statements of Japanese energy policy. It maintained the traditional goals, the so-called '3Es': energy security, environmental sustainability and economic efficiency. In addition, it reiterated two other goals that have been associated with Japan's controversial nuclear power program: safety and public understanding. In several other important respects, however, the 2010 BEP represented a departure from past policy. It included for the first time two other goals: the use of energy policy to promote more general economic growth and the need to restructure the energy industry. In addition, the 2010 BEP placed much

2 The energy self-sufficiency (ESF) ratio is the percentage of Japan's primary energy supply, which is produced domestically and consists primarily of renewable energy sources and nuclear power, since Japan produces only very small amounts of coal, natural gas, crude oil, and liquefied petroleum gases (LPG). The self-developed fossil fuel supply (SFFS) ratio is the percentage of imported coal, natural gas, oil, and LPG, which is produced by Japanese companies. The energy independence (EI) ratio is the percentage of Japan's primary energy supply, which consists of either energy produced domestically or imported fossil fuels, which are produced by Japanese companies.

more emphasis on combating climate change than did the NNES, which was primarily concerned with energy security. It is likely that this apparent obsession was in response to former Prime Minister Yukio Hatoyama's 2009 pledge to reduce Japan's greenhouse gas emissions substantially by 2020 (Duffield and Woodall 2011: 3745).

Post-Fukushima Developments

On the eve of Fukushima, Japan's energy security problems were clear, but the proposed solutions seemed to rely on a large degree of wishful thinking. Duffield and Woodall (2011) argue that achievement of many of the 2010 BEP's targets was likely to be challenging, even before the Fukushima disaster. Following on from the Fukushima nuclear accident, the Japanese government has been in the throes of reviewing its energy policy. Japan declared in an energy white paper in October 2011 that it would aim to reduce the dependency on nuclear power and revise the Basic Energy Plan, starting 'from a blank slate' (METI 2011a: 2). The former Prime Minister Naoto Kan announced that the government would have to 'start from scratch' in devising a new energy policy for the country. He has announced a major energy policy review that would promote solar and other alternative energies, stating that Japan should increase the share of renewable energy in power generation to 20 per cent by the early 2020s (Johnston 2011). Upon Kan's insistence and as a condition for his resignation as Prime Minister, the extended feed-in-tariff (FIT) that offers generous rates for renewable electricity has been the only major legislation passed in the aftermath of the Fukushima disaster. The Diet passed the extended FIT on 26 August 2011 (the *Act on Purchase of Renewable Energy Sourced Electricity by Electric Utilities*, METI 2011b) and the new FIT has been in effect since July 2012, encompassing generous electricity rates for photovoltaic (PV), wind power, small hydroelectric, geothermal and biomass (Ayoub and Yuji 2012, IEEJ 2011f: 5).

One of the guiding principles of government advocated by the DPJ administration was that the politicians had to wrestle power away from the bureaucrats while simultaneously reviewing Japan's energy policy. The bureaucrats were perceived as too powerful and influential in policy circles, *de facto* driving Japan's energy policy. Specifically, The Energy and Environment Council (Enecan), established under the National Strategy Council, was positioned at the heart of the policy debate, which was placed under the direct control of the Prime Minister. The council was tasked with formulating a set of basic principles on energy policy by amalgamating various policies discussed in the Advisory Committee for Natural Resources and Energy, the Central Environment Council, the Atomic Energy Commission and other pre-existing organizations (Miyamoto et al. 2012: 28). Enecan's *Innovative Strategy for Energy and the Environment*, formulated in September 2012, has set ambitious goals for energy conservation and electricity savings, cutting energy consumption by 19 per cent and electricity consumption by 10 per cent of 2010 levels by 2030 (IEEJ 2013a: 13). Further to public discussion of these options, Enecan was to present an outline strategy, or the new Basic Energy Plan, in late summer, but this did not materialise. With the LDP's electoral victory in December 2012, Enecan has become defunct and the ambitious goals that were set are likely to be reformulated as part of the new government's future energy policy.

Prior to the December 2012 election in Japan, Yoshihiko Noda's DPJ government, was set to reduce nuclear power dependency as much as possible (METI 2011a) and, in September 2012, following Enecan's recommendations, even hinted at phasing out nuclear power by 2040 (Dickie 2012). Although many expected that nuclear power would feature prominently in the pre-election campaign, it did not become a major campaign issue. A shift in voter perceptions regarding the future role of nuclear power did not materially impact the political arena (Scalise 2012a: 148). Although all parties other than the LDP pledged – albeit to varying degrees – to do away with nuclear power generation, the election outcome suggests that a large portion of unaffiliated voters who support an end to nuclear power voted for the LDP.

The newly elected government under Shinzo Abe's LDP is unlikely to commit to such a path. Shinzo Abe stated that nuclear phase-out is unrealistic and irresponsible (*ABC News* 2012). At the same time, the Abe government has adopted a cautious approach to energy policy. Japan's new energy policy is to be determined in the next decade and decision on reactor restarts in three years. The government will allow restarting nuclear reactors that have been deemed safe by the newly formed independent regulator, the Nuclear Regulation Authority (NRA). Initially, the NRA, which was established in September 2012, announced that it will decide the outline of the new safety standards by January 2013 and, following necessary adjustments, will draw up the draft new standards, the revised nuclear disaster preparedness guidelines and the *Regional Plan for Disaster Prevention* by March 2013. The new safety standards were finalized in July 2013, and the NRA started reviewing the power plants for restarting based on the new safety standards. In November 2013, the NRA announced that it has no fixed schedule to complete safety checks at idled nuclear power plants, possibly delaying reactor restarts and the supply of cheaper energy the Abe government wants to drive economic growth. In addition, the new METI Minister Toshimitsu Motegi stated that once experts' opinions were collected, the new government would take 'a major political decision' on whether or not to allow construction of nine reactors that currently exist only at the planning stage (cited in *The Asahi Shimbun* 2012).

According to Miyamoto et al. (2012: 29), Japan's future energy policy is likely to be formulated by placing priority on ensuring public safety and emphasizing sustainability and public trust, demand-side measures and utilization of diverse power and energy sources. More specifically, the basic direction of development of the energy mix is likely to be based on strengthening energy and electricity conservation; accelerated development and use of renewable energy to the maximum possible extent; effective utilization of fossil fuels (i.e., environmentally friendly use of fossil fuels), including a shift to greater use of natural gas; and reduced dependency on nuclear power wherever possible.

Conclusion

Most of Japan's energy security challenges are not new phenomena and have existed for decades. Consequently, there is historical contingency related to most of these challenges. This chapter places Japan's energy security challenges in the appropriate context and sets the scene for subsequent chapters to proceed with effective analysis and prescriptive advice. It examined the historical evolution of Japan's energy policy and energy security strategy during four major periods: the post-war (1950–1973), the oil crises (1973–1979), diversification away from oil, energy efficiency and reducing greenhouse gas emissions (1980–2011) and post-Fukushima developments.

Japanese national energy policy is a fragile consensus that unravels once the underlying assumptions surrounding the policy's purpose change. External shocks tend to provoke crises that force decision-makers to import workable blueprints of sector reorganization (Scalise 2012a: 151). Consequently, for most of the post-war era, Japan lacked an overarching energy plan or strategy. Instead, Japan's energy policies have been responsive to the country's needs and can best be described as reactionary. Beginning in 1967, the government published every two to five years a *Long-Term Energy Supply and Demand Outlook*, which forecast such important indices as energy demand by sector, primary energy supply by fuel and, in more recent years, energy-derived CO_2 emissions based on different sets of assumptions about the policies likely to be in place (IEA 2008: 21). However, the Outlook itself did not contain or determine policy. Instead, Japanese energy policy consisted of a patchwork of laws, regulations and programs, many of which were adopted in response to the oil crises of the 1970s. Policy-makers implemented *ad hoc* rules, regulations and laws in response to the oil shocks of the 1970s, the 'lost decade' of the 1990s and the global warming initiatives of the new millennium (Scalise 2012a: 141). It was only with the passing of *Basic Act on Energy Policy* in 2002 that the government created a more systematic and comprehensive energy policy planning structure, which resulted in three BEPs (2003, 2006 and 2010) and a NNES (2006).

Energy security has been a prominent theme in Japan's national energy strategy (Gasparatos and Gadda 2009: 4039). It is a continuing concern for Japan, a country that is heavily reliant on imported energy sources. The ability to be able to secure adequate access to energy imports at reasonable prices to satisfy the needs of the economy has been considered as a major determinant of Japan's overall security position. It has had profound implications for the health of the Japanese socio-economy and the political structures underpinning that economy. Yet, the risks of disruptions in energy markets are many. They include politically and market induced disruptions as well as accidents, such as at Fukushima. There is little predictability in the probabilities of these risks, their possible interactions and their cumulative effects.

Japan's energy strategies reflect its unique social decision-making system and position as a major actor in world energy markets, with few natural resources except for a large population and a system disciplined to pursue national economic security objectives. Japan's dependence on imported energy has had pervasive influence on strategic energy planning. Lacking significant domestic fossil fuel resources, since the 1973 oil crisis, Japanese energy policy has been centred on concerns of energy security. With the exception of Italy, which is securely linked into the EU energy network, no other industrialized nation has such a precarious dependence on other nations for energy supply. Any global disruption to energy supplies would have a greater impact on Japan than on any other nation (ANRE 2006). Cognizant of the risks that high levels of energy imports pose, a consistent tenet of Japan's national energy planners has been to wean the nation from high level of dependence on foreign energy. Securing stable energy supplies from abroad and establishing the 'best mix' of fuels and technology at home have preoccupied Japanese policymakers since the 1970s. At the same time, Japanese energy policy has been resilient to short-term fluctuations in energy prices. This suggests a high degree of stability in the fundamentals of energy policy even if there are short-term movements in energy prices.

Historically, the Japanese government has endeavoured to enhance economic security by minimizing energy costs, national energy security by reducing dependence on imported energy and environmental security by supporting sustainable energy solutions, which will not adversely impact the environment (METI 2006a). There has been a degree of symbiotic interplay between economic security, national energy security and environmental security policy objectives. A healthy economy can finance national energy security initiatives. Conversely, bolstering national energy security helps stabilize energy costs, which enhance economic security. Consequently, a synthesis has been aimed at by seeking policies that facilitate a long-term stabilization of energy costs at the lowest possible level. One approach to stabilizing energy costs was to endeavour to replace energy imports from politically unstable nations with those from stable nations. Another approach was to improve domestic electricity generation capacities (nuclear, renewable, etc.), which reduce reliance on imported energy and enhance environmental security. A final approach has been to improve energy efficiency in order to reduce reliance on imported energy while not having a negative effect on the economy. During the past decade, these energy policies were conceptually recast as the 3Es: (1) economic growth, (2) energy security and (3) environmental protection (IEA 2008). The Fukushima disaster challenged Japan's energy priorities and the absence of post-Fukushima policy commitments indicate that there is a hiatus regarding the country's future energy policy direction.

Chapter 3
Oil

Introduction

In his 1975 novel *Yudan*, Taichi Sakaiya, the former chief of Japan's Economic Planning Agency, paints a terrifying picture of what happens when conflict in the Middle East results in a 30 per cent cut in oil supply to Japan. Entire industries go bankrupt. Three million people die of hunger or fire. More than 70 per cent of the country's assets are abandoned or destroyed in riots. While the oil shocks that shook Japan in real life first in 1973 and once again in 1979 did not wreak as much havoc on society as depicted in *Yudan*, their impact on the lives of ordinary Japanese remains deeply etched on people's minds. Even today, the term 'oil shock' conjures up images of shoppers rioting to buy essential goods and memories of the sharp increases in consumer prices, which in some cases rose by as much as 20 per cent. These oil shocks caused the Japanese economy to record negative growth rates for the first time in its post-war history (Koike 2006: 44).

Japanese economy and foreign policy have always been constrained by the security issues posed by oil supply. For instance, the outbreak of the Pacific War (1941–1945) and the subsequent defeat, the first recession since the end of the war (1974) and the ongoing territorial disputes with neighbouring countries relate to the issues of oil shortage, and are thus firmly established in the mind of the government. Consequently, it is logical to begin analysing Japan's energy security with the focus on oil. Given that Japan is resource-poor and import dependent, it is not surprising that concerns over oil supply security have historically dominated energy policy debates in Japan.

The first section of this chapter surveys the evolution of Japan's oil policy. Given that oil has been the most securitized energy source over the past four decades, the chapter devotes significant attention to this survey. This is in contrast to other chapters on Japan's energy sources where historical policy evolution is devoted a less detailed attention. The second and third sections briefly outline sector organization and historical oil supply–demand balance. Overall, the first three sections provide background analysis for the remainder of the chapter. The discussion proceeds by examining major challenges related to Japan's oil supplies. The first challenge is Japan's continued pursuit of self-developed, or equity, oil, which has historically been an expensive policy, mired with inefficiencies in Japan's oil companies, with Japan never achieving desired policy goals. A related challenge is the zero-sum competition among Asian oil importers in which Japan has played an important role since the emergence of China as a major oil importer in the early 1990s. In fact, these three interrelated challenges stem from the

Japanese government's treatment of oil as a strategic commodity, which is not surprising given that Japan remains almost exclusively reliant on oil imports and the country's experiences during the 1970s oil crises. Despite Japan's success in reducing oil demand and the share of oil in the overall energy demand, the final challenge discussed in this chapter pertains to Japan's continued overreliance on the Middle East. Japan's international oil policy has suffered a number of policy failures despite being a key component of Japanese energy security strategy since the oil crises. Equity oil has only ever provided half the oil expected by MITI/METI and Japan remains largely dependent on the Middle Eastern oil.

Evolution of Policy

By 1952, when Japan regained its political independence, MITI had emerged as the agency entrusted with the industrial development of post-war Japan. During the 1950s, MITI promoted Japan's push toward heavy and chemical industrial development. Heavy energy- and oil-consuming industries, such as steel, non-ferrous metals, petrochemicals and synthetic fibres, came to lead the industrial growth of Japan, especially after the mid-1950s. Thus, the question of procuring sufficient quantities of oil at as low a price as possible for Japanese industrial users confronted MITI officials. In order to reduce the dependency of Japanese industrial users on oil products from Japanese subsidiaries of foreign oil firms, MITI came to favour Japanese oil refiners and marketers that were not in a dependent relationship. MITI suspected that foreign oil subsidiaries were not to be relied upon for cooperation in achieving the industrial goals of Japan. MITI's control over foreign exchange allotments for the import of vital natural resources, including oil, provided the government with an effective tool for regulating Japanese refining and marketing activities. A 'rule of thumb' used by MITI was to structure the oil industry in such a way that independent Japanese oil interests should constitute at least 30 per cent of the market share in both the refining and wholesale stages of the oil industry (Tsurumi 1975: 115).

While coal was the major fuel in Japan for decades, in the early 1960s concerns with the price of energy led to a rapid fuel substitution to more economic oil. In 1963, Japan's oil imports reached the level of 50 per cent of its total energy needs (Morse 1981b: 5). As the world oil market became increasingly glutted in the late 1950s and early 1960s, MITI's concern over procurement of crude oil dissipated. Yet, the government became concerned with the rapidly increasing dependence upon petroleum brought about by import liberalization and the concentration of suppliers, mainly among the international oil companies (Nemetz et al. 1984/1985: 555). The only goal remaining was to keep the prices of refined petroleum products in Japan as low as possible for the benefit of the heavy industries of Japan. For this purpose, MITI needed a new controlling tool. Its import licensing power was losing its regulatory effect over Japanese and foreign oil firms as various import restrictions were being eased in 1962 under mounting pressures from other countries, notably the United States. At this juncture, MITI promoted new regulatory legislation, the *Petroleum Industry Law* of 1962 ('Oil Act'), which would empower it to ration permits for the refining and sales operations of domestic oil firms. MITI argued that without the Oil Act it would not be able to help Japanese oil firms to compete with foreign subsidiaries. The ruling party, the LDP and business circles in Japan were easily persuaded to endorse the Oil Act, partly because they feared the consequences of relaxing import restrictions and the spectre of foreign domination.

Under the Oil Act, MITI was given authority to licence entry into refining, approve production plans and set standard prices. MITI used this authority to encourage Japanese firms to engage in refining and to discourage any domination of the industry by oil majors or the general trading companies. The resulting fragmentation of the refining industry was supposed to keep domestic petroleum product prices low (Nye 1981: 213–14). This legislation also required MITI's approval of any new refinery installation and expansion programs (Nemetz et al. 1984/1985: 555). According to Scalise (2004: 166), the Oil Act was written to achieve a stable supply of oil by controlling downstream oil refining, effectively authorizing the separation of upstream and downstream activities through permits to establish refineries or to purchase refining equipment. Several laws were subsequently enacted to reinforce the 1962 framework; among others, these include the *Japanese National Oil Corporation Law* (1967), the *Petroleum Supply and Demand Optimization Law* (1973) and the *Petroleum Stockpiling Law* (1975). As a by-product, these laws indirectly managed to protect the fragmented structure of the industry, thereby propping up many of the smaller firms that were dependent on existing regulations to stay in business.

The major thrust of MITI's oil policy during the 1960s was to let a limited number of large steel and chemical firms play one oil company off against another to obtain the lowest possible prices. For that purpose, MITI kept the structure of the Japanese oil industry fragmented by various means, including permitting new entries into both the refining and wholesale stages of the oil business and deliberately keeping the refining capacities out of balance with wholesale abilities in order to impede the vertical integration of refining and wholesale operations. In short, MITI counted on a greatly fragmented oil industry in Japan to engage in price competition in the sale of petroleum products, which would then be shielded from import competition by high tariffs on such products from other countries (Tsurumi 1975: 116). As the oil glut continued into the second half of the 1960s, MITI's initial interest in keeping down the number of refiners in Japan disappeared. Crude oil was readily available and MITI increasingly succumbed to political pressure from petrochemical firms, trading companies, public utilities and other business interests that wanted to enter the oil industry without worrying about their ability to procure crude oil (Tsurumi 1975: 117).

The proliferation of oil firms in Japan during the late 1960s was limited to downstream operations, i.e., the refining to retailing stages. None of Japan's refiners ventured abroad for crude oil production. Financially too weak to undertake risky ventures of crude oil exploration, they were content to count on the worldwide glut situation to ensure a ready supply of crude oil. By 1973, major foreign oil companies had come to supply over 80 per cent of Japan's needs for crude oil; the only Japanese overseas producer of crude oil, the Arabian Oil Company, supplied less than 9 per cent. Trading companies procured the remaining 11 per cent from independent foreign oil firms (Tsurumi 1975: 118).

While economic realities moved Japan to adopt a pragmatic but generally myopic energy policy during the 1960s, the quest for long-range security did not stop. The realization that petroleum was the major source of Japanese energy supply and that demand growth could not be controlled under the prevailing conditions, prompted attention to ensuring the stability of supply. In February 1967, a report by MITI's Advisory Committee on Energy recommended a more intensive thrust in this direction. One initiative in this direction was the relatively small investment by the Japanese government in 1967, incorporating the Japan Petroleum Exploration Company (JAPEX), into the government-owned Japan Petroleum Development Corporation (JPDC) in 1967 to encourage Japanese private-sector investors to participate in overseas petroleum exploration and development, as discussed in more detail below. As a consequence, between 1968 and 1973 the number of Japanese firms engaged in overseas exploration increased from 8 to 49, the number of wells drilled from 44 to 180 and expenditures on exploration and development from $44 million to $488 million (Vernon 1983: 96).

The oil crisis in 1973, triggered by the Yom Kippur War, found Japan in a highly vulnerable position. A reduction by 17 per cent in the volume of world oil trade meant an 8 per cent reduction in Japan's supply. The actual interruption in Japan's physical access to energy was augmented shortly thereafter by a threat to its economic access. The price of crude oil increased four times within a few months, more than tripling in its level. The energy-intensive industrial structure of Japan presented a major problem, with the dominant share of production being in iron and steel, non-ferrous metals, petrochemicals and other chemicals, automobiles, plastics, synthetic fibres, machinery and ships. The response to the November 1973 cutbacks included the imposition of emergency energy conservation programs. Large companies in eleven industrial sectors were required to reduce their purchases of oil by 10 per cent below the October level. In January 1974, when the delayed impact of the cutback was felt, the reduction in industrial oil quotas was raised to 15 per cent. By March 1974 the regular oil flow to Japan had been restored, but the economic problems that the new oil prices were creating in the country, which was already in the midst of a growing inflation, seemed serious. From a decade characterized by two-digit annual growth in GDP, Japan entered a period of sustained growth but at average annual rates about half those experienced over the previous decade (Nemetz et al. 1984/1985: 557).

In 1973, the Japanese economy was highly dependent on imported oil. After the 1973 oil price shock vigorous policy action was undertaken by the Japanese government in an attempt to reduce Japan's energy consumption and, in particular, its dependence on imported oil (Perkins 1994: 595). As discussed in Chapter 2, these policies included both direct attempts to influence the amounts and types of fuels used throughout the economy and industry structure policies aimed at facilitating a shift out of energy intensive industries.

Japan's problem of oil import dependence and the need to improve security of supply has led to the development of a sophisticated 'resource diplomacy' after the first oil crisis (Caldwell 1981). Prior to the 1973 oil crisis Japan imported approximately 40 per cent of its oil needs from the Arab members of the OPEC. Understandably, therefore, when the Organization of Arab Petroleum Exporting Countries (OAPEC) cut its oil production by 25 per cent in November 1973 and classified Japan as an 'unfriendly' state, Tokyo was thrown into a state of panic. Moreover, given that a few Western majors controlled the bulk of Japan's oil supply, Japanese industry feared that the foreign majors would favour American and other Western customers over Japan for the duration of the crisis. The panicking of Japanese industries worsened the overall situation by driving oil prices up further (Morse 1982: 261).

In order to restore interrupted supplies, Japan shifted its foreign policy posture and abandoned the principle on which its post-war external relations had been based – the separation of economic and political matters (Wu 1977). Japan had come to a realization that because of its dependence on Arab oil producers, political relations with these countries could not be ignored; that the United States could not on account of its links with Israel be relied upon to protect Japanese oil interests; and that the American-based oil majors (which together with the Europeans provided Japan with nearly 80 per cent of its crude oil) might be diverting oil to the United States at Japan's expense. The severe shortages suffered by Japan when the international companies diverted Iranian and Indonesian oil to the United States and other markets brought the realization that in the event of a serious crisis the US companies would probably give preference to US interests (Sinha 1974: 342). Believing its survival to be at stake, Japan felt compelled to distance itself from the United States and to opt for an independent diplomatic posture – one which reflected more closely the Arab point of view as well as Japan's political commitments to the Middle East, but one that did not diverge to an unacceptable degree from that of the United States (Tsurumi 1975: 126, Yorke 1980: 434, Lam 2009: 117).

As a result of Japan's resource diplomacy following the 1973 oil crisis, the share of oil imports purchased under bilateral deals between Japan and foreign governments increased from 5.4 per cent in 1972 to 31.4 per cent in 1978 (Morse 1982: 261). However, up to the autumn of 1978, international oil companies still provided approximately 70 per cent of Japan's supplies. When those companies lost access to crude oil supplies in Iran and Nigeria, they were forced to cut back on sales to third parties in order to supply their own affiliates and stopped supplying oil to non-affiliated companies in Japan. Since Japan had discouraged large refining efforts by the majors, many of these third parties were independent Japanese refiners who felt a sense of panic. Japan's response was twofold. First, encouraged by MITI, Japanese trading companies doubled the amount of oil bought directly from the oil producers or on the Rotterdam spot market. Stockpiling to the limits of their capacities, Japanese traders were held responsible for driving up spot prices (Morse 1982: 262). The Japanese firms paid extreme spot market prices for what oil they could get and submitted to extraordinary contract terms. By the end of 1979, direct deal and government-to-government oil accounted for 45–50 per cent of total imports, thus significantly reducing the oil majors' control of Japanese oil supply (Yorke 1980: 441). By 1980, the share of international oil companies in Japan's supplies further dropped to 44 per cent (Nye 1981: 216, Morse 1982: 261). Second, as continuation of broader resource diplomacy, Japan also accelerated efforts to strengthen bilateral links at economic and political levels with the Arab oil producers. With the sale of crude oil linked to the acquisition of Japanese technology through inter-governmental and corporate co-operation, Japan's trade prospects blossomed.

The drastic drop in the oil prices during the mid-1980s helped in changing the people's view of oil as a market commodity, rather than a strategic resource. The ample production and high liquidity of oil in the developed international oil market led to oil being treated as a general commodity, which was relatively easy to procure from the market. This implied that public and policy leaders in Japan might not have considered the priority of oil security as a serious matter. Since the mid-1980s, Japan gradually eased regulations and laws in order to bend to public pressures to reduce high gasoline, kerosene and naphtha prices in a domestic market that was experiencing a clear slowdown in economic growth compared to the previous two decades. Technically, such liberalization of the petroleum industry began with the 1986 *Refined Petroleum Import Law*. The MITI eased its requirement for obtaining a license to import refined products, answering calls from domestic and international actors to open markets. However, incumbent oil refiners successfully lobbied to limit the impact of the new legislation by requiring only those refineries in Japan with existing domestic production, stockpiling and quality control to import oil products. The requirement effectively blocked all imports from non-incumbent players, making 'petroleum liberalization' a paper tiger (Scalise 2004: 166–67).

In March 1996, the *Refined Petroleum Import Law* expired and the *Gasoline Sales Law* and *Stockpiling Law* were also revised. Emphasis shifted away from 'energy stability' of imported petroleum to 'energy efficiency', partly aided by advocates of reform in the newly established Hosokawa Morihiro coalition government. Extreme inefficiencies and ineffective legislation kept gasoline prices high in the 1980s, but real liberalization of the market led to greater market competition and increased consumer welfare gains despite yen depreciation and stable crude oil prices. The key to such consumer welfare gains emanated from laws enacted to stir competition in the mid-1990s. With the *Refined Petroleum Import Law* removed, stringent restrictions on petroleum imports were also removed; import licensing was readjusted to require new capacity to decrease from 120 days' worth of product storage (a clear structural barrier to market entry) to only 60 days' worth of product. In turn, the abolition of the Specified District System, which prohibited new entrants from targeting fragmented districts in which there were already a number of gasoline stations, allowed for greater competition among retail players. Consequently, incumbent gasoline stations began to lower prices in an effort to build customer loyalty. Corporate profit margins, historically low by world standards, continued to decrease in the newly competitive operating environment. Consequently, consolidation, asset streamlining and cost cutting were initiated in the industry (Oyama 2000: 16–17). In 1999, Nippon Oil and Mitsubishi Oil, for example, merged to form the largest Japanese oil distributor by market capitalization (Scalise 2004: 167).

The first BEP (2003) expected oil to play an important role in Japanese energy supply, which was seen as a point of departure from the past 'policy away from oil' (Yokobori 2005: 316). It argued for enhanced efforts to secure a stable oil supply. In 2006, the NNES was developed in response to renewed concerns about Japan's energy security due to rising oil prices, a revival of resource nationalism and growing regional competition over energy resources, particularly with China (Atsumi 2007: 29, Duffield and Woodall 2011: 3742). Thus, the NNES placed primary emphasis on and sought to bring greater attention to the issue of oil security. Against this backdrop, Japan's 2006 NNES (METI 2006a) set forth three specific numerical targets related to oil for Japan to achieve by 2030:

- the dependency on oil, as a percentage of the total primary energy supply, should be reduced to 40 per cent or less from the present level of a little less than 50 per cent;
- the percentage of oil in total energy consumption by the transport sector should be reduced to about 80 per cent from the present level of almost 100 per cent; and
- the percentage of equity oil secured by Japanese companies in the total supply of imported oil should be increased to 40 per cent from the present level of 15 per cent.

During this time, the policy-planners and decision-makers were moving towards general agreement that Japan should also enter into the scramble for oil (Atsumi 2007: 30). During the following years, however, more of a balance was restored in Japanese energy policy. Concerns about security of supply abated somewhat, despite a continued increase in oil prices, while concerns about environmental sustainability, especially climate change, returned to the fore. In recent years, Japan's policy has remained committed to lowering Japan's oil dependence with the ban on construction of new oil-fired power plants still in place and the government's commitment to reducing the share of oil in the transportation sector and substituting for oil in the industrial sector (Petroleum Association of Japan 2012: 7).

Sector Organization

Japan has 4.479 million bpd of oil refining capacity at 27 facilities and has the second largest refining capacity in the Asia–Pacific region after China. JX

Nippon is the largest oil refinery company in Japan and operates seven refineries with 1.24 million bpd of capacity (Petroleum Association of Japan 2012: 69). In recent years, the refining sector in Japan has been characterized by overcapacity since domestic petroleum product consumption has declined due to the contraction in industrial output and the decline in transportation fuel demand because of mandatory blending with ethanol. For example, refinery utilization rates have remained below 80 per cent since 2008, with only 74.2 per cent utilized in 2011, the lowest in the past two decades (Petroleum Association of Japan 2012: 9). Consequently, Japan scaled back refining capacity by 850,000 bpd since 2000. In addition to declining domestic demand, Japanese refiners now must compete with new state-of-the-art refineries in emerging Asian markets. The total refining capacity per refinery is relatively small compared to refineries in other Asian countries, particularly the new refineries in China, Singapore and India. Private Japanese firms dominate the country's large and competitive downstream sector, as foreign companies have historically faced regulatory restrictions. But over the last several years, these regulations have been eased, which has led to increased competition in the petroleum refining sector, with Royal Dutch/Shell and ExxonMobil operating refineries in Japan (EIA 2012b).

The Japanese government is seeking to promote operational efficiency and in 2010 METI announced an ordinance that would raise the cracking to crude distillation capacity ratio that refiners had to meet by March 2014 from 10 per cent to 13 per cent or higher. This ordinance is intended to increase refinery competitiveness within the country and will likely lead to further refinery closures if implemented. According to IEEJ (2013a: 7), domestic oil demand is forecast to drop in the future and numerous refineries will be closed by March 2014. *FACTS Global Energy* anticipates that the ordinance could remove an additional 600,000 to 800,000 bpd of refining capacity as companies rationalize their expenditures. Announced closures along with the METI legislation could lower refining capacity by a total of 1.3 million bpd by 2014 (EIA 2012b). For example, JX Nippon aims to shut down 600,000 bpd of capacity between 2008 and 2015.

Japan's oil stocks are well in excess of the IEA's 90 day equivalent net import requirements. As of February 2013, the country held the equivalent of 163 days of net imports, including state-owned and private-sector stocks (IEA 2013b). Japan's stockpiles (public and private) are far greater than in most industrial countries. By maintaining one of the largest strategic petroleum reserves in the world, Japan, in coordination with other International Energy Agency (IEA) member states, can mitigate the effects of global oil supply disruptions. Private refiners in Japan are required to maintain petroleum product stocks equivalent to at least 70 days of imports, which imposes large additional costs to these companies. This regulation was relaxed to 67 days after the Fukushima incident. Nevertheless, in February 2013, private refiners held 72 days, with the government holding 91 days' worth of imports (IEA 2013b).

Japan has very limited domestic oil reserves, amounting to 44 million barrels as of January 2012, down from the 58 million barrels in 2007. Japan's proved domestic reserves would supply the country for ten days at current rates of consumption. In 2011, Japan's total oil production was roughly 130,000 bpd, of which only 5,000 bpd was crude oil. The vast majority of Japan's oil production comes in the form of refinery gain, resulting from the country's large petroleum refining sector (EIA 2012b). While Japan has maintained its domestic oil production, the output has never been sufficient to meet the demands of the world's third biggest consumer. Thus, Japan has had no choice but to depend on imported oil for almost all of its domestic demand.

Although Japan is a minor oil producing country, it has a robust oil sector comprised of various state-run, private and foreign companies. JAPEX was founded in 1955 as a government-owned company in order to seek petroleum self-sufficiency for Japan. In 1965, JAPEX's operational range moved overseas and, in 1967, JAPEX was incorporated into the government-owned JPDC and effectively nationalized. Since there is no state-owned oil company in Japan, the JAPEX's role was to form joint investments with private oil companies for policy implementation. JAPEX was tasked with stimulating overseas petroleum development and its duties were progressively expanded in order to pursue energy diversification and energy security. More specifically, JAPEX was tasked with providing equity capital, loans and loan guarantees for overseas oil exploration, as well as technical guidance to the private sector (Nemetz et al. 1984/1985: 556). In 1972 the provision of credit for private overseas energy projects became central to these activities. Later the corporation also began providing funds in the form of equity capital and loans, to assist firms engaged in downstream refining and marketing (Vernon 1983: 95). JAPEX was modelled on the French and Italian oil firms (ELF and ENI, respectively) and was initially tasked with providing capital for projects using revenues collected from oil consumption taxes (Samuels 1987: 208–9). This arrangement was meant to ensure a stable flow of energy through the development of oil and gas resources. In 1970, JAPEX was separated from the JPDC and reorganized as a privately owned company, but continued operating under strict government guidelines.

In 1978, the JPDC changed its name to the Japan National Oil Corporation (JNOC). The JNOC took JPDC's roles of promoting oil exploration and production domestically and overseas, but also charged with the operation of oil stockpiles to its activities. Until 2002, Japan's oil sector was dominated by the JNOC. Japan was very aggressive in its search for overseas oil supplies in the 1970s. This quest started in the mid-1960s, but grew rapidly after the oil crises. During this time, Japan was concerned about dependence on the major international oil companies for over 40 per cent of its oil imports. In 1979, Exxon, BP and Shell decided to cut back shipments to Japanese oil refineries as a result of the dispute between Iran and the United States. It was believed in Tokyo that in a crisis, these foreign importers could channel oil away from Japan (Yorke 1980: 433). JNOC often made a 50:50 share arrangements with the private companies in overseas petroleum development. In dealing with projects that the private companies were reluctant to be involved in, but MITI wished to pursue, JNOC could provide 100 per cent of initial investment and bear the full cost if things did not proceed according to initial plans. JNOC contributed greatly to the building of substantial strategic petroleum stockpiles in Japan (Liao 2009: 111). JNOC built up a portfolio of 119 projects before the company was taken apart by the government in 2002 as a consequence of its poor performance in terms of debt accumulation and inability to achieve self-developed targets, as discussed below (Koike et al. 2008).

As oil prices have risen over the past decade, a resurgence of fears over security of supply in Japan triggered the creation of the Japan Oil, Gas and Metals National Corporation (JOGMEC) from the remains of JNOC in July 2002, with changes taking effect in February 2004, pursuant to the *Law Concerning the Japan Oil, Gas and Metals National Corporation.* Many of JNOC's activities were taken over by JOGMEC, which was charged with aiding Japanese companies involved in exploration and production overseas and promoting commodity stockpiling domestically. JOGMEC was established to perform three main roles in supporting private companies: financial support, research and development for technology and oil stockpiling. JOGMEC's financial support covered not only projects in unexplored areas but also acquisition of existing fields. Yet, the extent of JOGMEC financial contributions was narrowed and scaled down to investment and guarantee in relation to borrowing up to 50 per cent, as opposed to 70 per cent, of the total exploration costs. In terms of its legal form, JOGMEC did not become a special public corporation as JNOC but was established as an incorporated administrative agency that did not enjoy the privilege of government guarantee for fund-raising or exemption from tax liability. Around the time when JOGMEC was created, there were few demands for a stronger national commitment by taking more direct risks in the debate over reviewing the role of the government in overseas oil development (Liao 2007: 33, Koike et al. 2008: 1772–73).

In the past decade, the government also promoted integration in the private sector of the oil-development industry. In 2003, the Advisory Committee for Energy under METI proposed that Japan needed a central firm for securing a stable overseas supply, because creating such a firm would improve its bargaining power, strengthen the company's finances, boost fund-raising and investment capacity and build up technology and human resources. In 2004, the government founded a new international company, Inpex, of which 18.9 per cent is owned by the METI, with the Minister holding one special class of shares. Aiming at a national flagship company, Inpex and Teikoku Oil entered into an administrative merger in April 2006. The merger by Japan's largest and third largest oil developers was completed in 2008, expanding overseas access to desirable projects by means of large-scale corporate resources. The merger was designed to increase the leverage of these companies against other players in the international oil market (Vivoda and Manicom 2011). Inpex and Teikoku Oil agreed to METI's proposal due to a sense of crisis in the harshly competitive global environment (Koike et al. 2008: 1773).

In 2006, the Japanese government launched the NNES and set for the newly created Inpex the objective of securing 40 per cent of the nation's oil imports through equity assets overseas (METI 2006a). The government also ensured the continuation of JAPEX, 34 per cent owned by the government (Koike et al. 2008). The new policy emphasized the need to address the country's rising dependence on the Middle East, the growing constraints on foreign direct investment (FDI) in oil-producing states and greater competition from other Asian oil importers such as China and India. Support in Japan for overseas investment was part of a wider strategic program of action, which included diversification of sources of supply, diplomacy, technological assistance and aid. A vigorous and open debate took place in Japan between those who preferred to rely on and support international markets and those who distrusted them and preferred a more autonomous approach, with the latter side prevailing (Herberg 2007). Consequently, the Japanese government still relies heavily on strategic approach to ensure its energy security. As part of this strategic approach, through JOGMEC, the Japanese government sets ambitious targets for Japanese oil companies pertaining to oil imports from overseas assets and supports the companies with government funds, bank loans and resource diplomacy.

Governmental strategy in Japan is motivated by the desire to build internationally competitive oil companies. Japan supports its companies as they are perceived to be latecomers and, as such, 'inferior in both capital and technologies' to American and European oil competitors (Petroleum Association of Japan 2012: 25). The weakness of Japanese oil companies results in part from their lack of domestic oil and gas production and the revenues and expertise that such production brings (Andrews-Speed 2012: 36). Japanese oil companies are involved in approximately 130 oil and gas development projects around the world (Petroleum Association of Japan 2012: 25). In 2010, Inpex had 71 projects in 26 countries (Andrews-Speed 2012: 33).[1] Japan's overseas oil projects are primarily located in the Middle East and Southeast Asia.

While Japanese oil companies are not state owned, they rely on state support beyond that provided by JOGMEC. According to MoFA (2004), 'Japan is working to strengthen relations with oil exporters and to improve the investment environment'. This aims to encourage Japanese companies to produce oil overseas and to build political relationships with oil-exporting countries. More recently, the state-backed Japan Bank for International Cooperation (JBIC) has focused on acquisitions of overseas resources rather than on development and has, according to *The Economist* (2009), $12 billion available. Japan seeks to purchase oil directly from supplier nations at reduced prices in exchange for investment or official development assistance (ODA).

1 Besides Inpex, other Japanese companies and trading houses involved in exploration and production projects overseas include: Cosmo Oil, Idemitsu Kosan, Japan Energy Development Corporation, JAPEX, Mitsubishi, Mitsui, Nippon Oil, and others. Many of these companies are involved in small-scale projects that were originally set up by JNOC. However, many are involved in high-profile upstream projects involving major investments in overseas ventures in recent years (EIA 2012b).

Supply–demand Structure

Between 2000 and the Fukushima disaster, oil demand in Japan has declined by 20 per cent (Figure 3.1). This decline stems from structural factors, such as fuel substitution, an aging population and a shrinking workforce, resulting in fewer vehicles on the road and government-mandated energy efficiency targets, which have resulted in higher fuel efficiency in the transportation sector. In addition to the shift to natural gas in the industrial sector, fuel substitution is occurring in the residential sector as high prices have decreased demand for kerosene in home heating. Japan consumes most of its oil in the transportation and industrial sectors. In the generation of electric power, the share accounted for by oil has dropped from around 75 per cent in 1973 to the low of 8 per cent in 2010 (EDMC 2010: 188, Petroleum Association of Japan 2012: 19). The share of oil in electric power generation increased to 14 per cent in 2011 as a consequence of the Fukushima disaster and the ensuing nuclear power shutdown. In the industry sector, the share of energy provided by oil declined from 61 to 43 per cent between 1973 and 2010. The transportation sector, however, remains entirely dependent on oil. Japan is also highly dependent on naphtha and low sulphur fuel oil imports. While demand for naphtha is falling as ethylene production is gradually being displaced by petrochemical production in other Asian countries, demand for low-sulphur fuel oil is increasing as it replaces nuclear electric power generation (EIA 2012b).

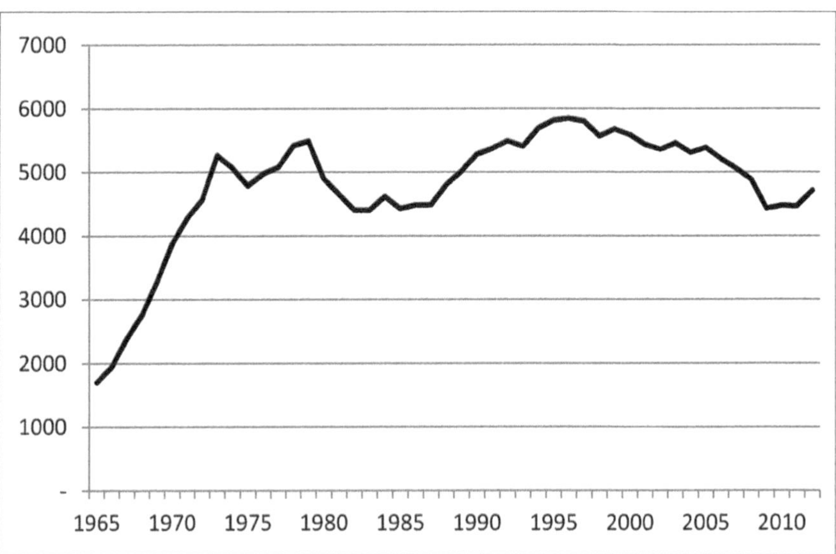

Figure 3.1 Japan's oil demand (1965–2012; thousand bpd)
Source: BP 2013.

Since the Fukushima disaster, Japan's oil consumption increased by approximately 240,000 bpd due to post-disaster reconstruction works and mainly substitution of crude oil and low sulphur fuel oil for the suspended nuclear power after the Fukushima incident (BP 2013). After the Fukushima incident, Japan has been increasing imports of crude oil for usage in thermal power plants. The March 2011 earthquake caused an immediate shutdown of six refineries with 1.4 million bpd or approximately 31 per cent of the total refining capacity. However, the country ramped up imports of refined products, particularly low sulphur fuel oil, in order to offset shortfalls in fuel supply for power generation until refineries were restored. In 2011, fuel oil imports surged to 102,000 bpd, rising from 58,000 bpd in 2010 while crude refining was down by 5.6 per cent to 3.4 million bpd in 2011, remaining at the same level in 2012 (EIA 2012b, BP 2013). As a consequence of the refinery shutdowns in recent years and subsequent increased demand for petroleum products, but also against the backdrop of relatively high oil prices, Japan's crude oil and petroleum product import cost increased from ¥11 trillion in 2010 to ¥14.7 trillion in 2012 (MoF 2013).

Japan's transportation sector is almost entirely dependent on petroleum products. In the NNES of 2006, the government articulated the goal of decreasing dependency on petroleum products in total energy consumption by the transportation sector to 80 per cent by 2030. In recent years, the government has been aiming to enhance security by lowering oil demand through the introduction of biofuels for transport. According to the *Petroleum Association of Japan*, the supply of biofuels in the transportation sector in 2010 was about 0.35 per cent of domestic gasoline consumption. While the biofuels industry in Japan is less developed than that in Brazil, the US and Europe, the Japanese government has announced a number of measures to accelerate use of bioethanol (E10). In 2008, the government introduced tax incentives to encourage the use of bioethanol by amending the *Act on the Quality Control of Gasoline and Other Fuels* implemented by METI. The gas tax is usually ¥53.8 per litre. Under the new tax system, if a fuel contains 3 per cent bioethanol, the gas tax is lowered by ¥1.6 per litre. In October 2008, the *Law to Promote the Usage of Biomass Resources to Produce Biofuels* came into force. The legislation includes tax breaks and financial assistance for biofuel manufacturers and farmers producing feedstock, such as agricultural cooperatives and private businesses. In 2012, the government permitted sales of E10 gasoline and vehicles designed to use E10. The aim is to increase bioethanol production from 50,000 kilo litres (kl) in 2011 to 6 million kl by 2030 – equivalent to 10 per cent of annual gasoline use in Japan (Iijima 2012). Due to the increase in food prices during the past few years, there is a broad debate within Japan about the use of food crops to produce biofuels.

Other means to reduce the dependence on petroleum products in the transportation sector is through the introduction of gas-to-liquids (GTL) conversion, batteries, hybrid vehicles, hydrogen, fuel cells and clean diesel. A Japanese consortium consisting of oil companies, oil refiners, engineering firms and JOGMEC in April 2012 announced a technological breakthrough in their research on the conversion of gas into synthetic fuels. Calling their research the Japan-GTL process, the consortium promises that the breakthrough will bring down the cost of GTL conversion. In the medium to longer-term, Japan is also aiming to build a new generation of efficient, electric vehicles to reduce demand for petroleum products in the transportation sector (Martin 2011). In addition, fuel cell vehicles may be commercialized by 2015 (Interview with Hideaki Fujii, 11 January 2013). In order to help reduce GHG emissions, Japanese auto industry is promoting 'clean energy vehicles', which include battery cars, hybrid and natural gas fuelled cars. Among those, hybrid cars are most prevalent in Japan, with the number exceeding 2 million in 2011. In 2012, a new standard for vehicle fleet efficiency was established for gasoline fuelled passenger vehicles. The standard has set a goal to attain vehicle fleet efficiency of 20.3 km per litre by 2020, compared to the 2009 level of 16.3 km per litre (Iijima 2012).

Nevertheless, given its dominance in the transportation sector, oil is expected to remain the most important energy source in Japan for a considerable time, even with progress in the development of alternative energy sources and technologies. According to Toichi (2006: 17), the 2006 targets for oil dependency ratio in the transportation sector will be difficult to achieve. Moreover, given that almost half of oil consumed in Japan is by the industry (i.e. as petrochemical feedstock), there is also much potential to substitute gas for oil (Interview with Akira Iishi, 16 January 2013). While oil demand is expected to drop due to the population decline and substitution in transportation and industry sectors, oil will retain the largest share of Japan's energy supply until at least 2030 (Petroleum Association of Japan 2012: 18).

Continued Pursuit of Self-developed Oil

Japan can never be fully free from the prospect of energy shortage. Consequently, particularly since the 1973 oil crisis, Tokyo has viewed developing oil overseas through Japanese companies as an integral part of import diversification policy and has relied on state institutions to support this policy (Nemetz et al. 1984/1985). Understanding the vulnerability caused by its high dependence on foreign producers from past experience, Japan has long made efforts to increase its self-developed oil production in overseas oil fields. The pursuit of equity oil through JNOC emerged as an extension of Japan's resource diplomacy following the oil shocks.

However, despite fully understanding the value of self-developed oil supplies through first-hand lessons and, having invested considerable effort in achieving various objectives, Japan has failed with its previous self-developed oil objectives. The first national target of securing self-developed, or *hinomaru*, oil supplies was set 40 years ago and was downwardly revised several times due to changes in the external environment by past administrations; however, none of the targets were achieved (see Table 3.1).

Table 3.1 Japan's self-developed oil targets and outcomes

Year of setting target	Target	Target year	Outcome
1965	30 per cent of total imports	1985	10.7 per cent
1978	1.5 million bpd	1990	0.45 million bpd
1983	1.2 million bpd	1995	0.69 million bpd
1993	1.2 million bpd	Early 2000s	0.58 million bpd
2000	Cancellation of numeric target as it promoted ineffective management		
2006	40 per cent of total imports	2030	TBD

Source: Koike et al. 2008.

In 1965, the amendment of the oil-development law allowed JAPEX to explore foreign reserves. In order to reduce the dependencies on imports and the Middle East for oil resources, in 1978, JPDC was transformed to JNOC and established as the national supporting organization for private companies. At that time, Japan secured only 14.2 per cent of total oil imports through its own companies, Arabia Oil Company (AOC) and North Sumatra Oil Development Cooperation Co. Ltd. (NOSODECO). AOC discovered the Khafji oilfield in 1960 and Hout oilfield in 1963, in the offshore Neutral Zone between Saudi Arabia and Kuwait. NOSODECO cooperated with Permina (now Pertamina) to redevelop the Rantau oilfield in Indonesia's Sumatra Island. The Japanese government decided to push overseas activities such as these projects by private companies. In 1967, the government set a target of obtaining 30 per cent of Japan's 1985 oil requirements from Japanese-developed sources, interpreted as finding oil wherever possible, including in the Middle East. The central aim of this policy was to reduce dependence on the international oil companies, rather than dependence on the Middle East (Surrey 1974: 216).

JNOC's support to private companies was mainly provided in the form of equity, loans and guarantee of liabilities. JNOC invested or loaned to private companies up to 70 per cent of total expense for the exploration stage of their overseas projects. If the project progressed to the next development stage, JNOC also provided a guarantee of obligation for up to 60 per cent of total loans from financial institutions for project companies. However, if the exploration failed or the fields discovered were found not to be feasible, the repayment obligation was waived and the project companies were liquidated. The Japanese government originally planned that JNOC should play a supportive role for the domestic downstream oil industry to arrange for the smooth delivery of oil supplies developed overseas by domestic companies for use in the domestic market. International oil companies dominated domestic oil delivery due to their long-term contract arrangements with the Japanese downstream industry and the smooth introduction of self-developed oil supplies was predicted to be very difficult. However, this plan was taken away by the domestic downstream industry through the *Petroleum Association of Japan*, which was concerned about possible government-imposed pressure to accept the non-commercial self-developed oil for national security reasons (Koike et al. 2008). As a result, the involvement of the Japanese government in overseas oil development provided exclusive indirect support to the upstream industry.

These aspects of governmental support rapidly increased the volume of overseas projects. During the first 20 years after the establishment of the JPDC, 119 projects were accepted by JNOC for support. Most of these projects ended up with a large amount of debt. In addition to the financial consequences, Japan's mixed effort, involving both the government and the private sector, failed to achieve any of the original targets of increasing the ratio of self-developed oil in terms of total imports and also the government's broader objective of diversification of the source countries for oil imports away from the Middle East, discussed later in this chapter.

A drop in oil prices during the second half of the 1980s and a serious and longstanding post-bubble stagnation in the 1990s left the public opinion opposed to the extravagant use of tax money, thus prioritizing economic efficiency over oil security. In particular, debate over ineffective fiscal investment and loans to public corporations along with the vested-interest structures attracted public attention, leading to some reform. During the 1990s, Japan's autonomous development policy came under increasing scrutiny. JNOC and the policy of promoting Japanese-developed oil supply also became subjects under reconsideration (Koike et al. 2008: 1772). Many felt that supporting domestic production of foreign oil had become too expensive. Low oil prices had made investment in oil reserves risky and unattractive for Japanese upstream corporations. Many resource projects in which the government-financed JNOC had invested were facing insolvency. JNOC drew on special government funds collected from oil consumption taxes to subsidize hydrocarbon exploration investment by Japanese oil companies. However, most of these investments were failures. Investigations undertaken in the late 1990s revealed that JNOC had accumulated ¥1.4 trillion ($9.7 billion) in bad debts and investments, accrued as a result of poorly planned and poorly executed large-scale oil exploration projects (Koike 2006: 45, Christoffels 2007: 17–18). Of the 112 Japanese upstream oil development companies in which JNOC invested, nearly 100 were found to be operating in the red and making little contribution toward increasing the flow of oil to Japan.

In the slipstream of liberalization policy, these investments in upstream oil development projects came under fire from politicians and business leaders alike. In November 1998, a former Minister of METI, Mitsuo Horiuchi, accused JNOC of the ineffective management of tax money. Being an experienced politician, Horiuchi ran his own company and was familiar with business accounting. He substantiated his views about the mismanagement of both the JNOC and its subsidiaries by reviewing a number of their financial statements and opposing government officials' explanations that the JNOC remained in surplus. Politicians and business likened JNOC-supported overseas ventures to 'white elephants'.

JNOC was accused of gross inefficiencies and bad risk management and critics referred to it as a 'retirement house' for METI bureaucrats (Hosoe 2005). Bowing to public anger, JNOC was dismantled in 2002 as part of the structural reform carried out by the administration of the Prime Minister at that time, Junichiro Koizumi. JNOC was replaced with JOGMEC and, to avoid a repeat performance, much more restrictive conditions were imposed on the ability of the new organization to subsidize upstream Japanese oil companies (Evans 2006: 20–21). Budgets for government support of resource projects were cut (Hosoe 2005). Japan's 1993 self-developed oil target was also withdrawn in 2000 as it promoted ineffective management.

One of the reasons for Japan's failure to achieve self-developed oil targets is that MITI/METI was never able to get private Japanese oil interests to cooperate or merge. Domestically they were unable to achieve upstream/downstream integration between Japanese oil companies. Overseas they were unable to get the companies to cooperate along national lines. For example, in Iran and Libya there were fiascos when the governments wanted to deal with a 'single Japanese entity', but METI could not get the Japanese companies to cooperate in effective consortia. Underlying these failures was also a lack of real will. Low oil prices made FDI in oil unattractive in the 1990s and many projects financed by JNOC incurred heavy losses (Christoffels 2007: 11).

Even when JNOC was still functioning, METI's ability to guide private industry's decision-making was limited. As a consequence, Japan's autonomous development policy, aimed at maximizing the amount of equity oil for Japan, was hardly a structured affair. METI's ability to coordinate Japanese upstream investments has been weakened by the forced break-up of JNOC. The fact that METI could engineer the merger of upstream oil companies Inpex and Teikoku in 2006 shows that the ministry has considerable sway over the part of the industry that was originally set up with government assistance. However, its influence over independent companies such as Idemitsu and Nippon Oil is much more limited. METI has lost credibility in the upstream oil and gas market because of the problems at JNOC and this has adversely affected METI's ability to cooperate on strategic energy policy with the private sector (Christoffels 2007: 49–50).

Historically, Japanese self-developed overseas ventures never generated more than 15 per cent of imports (Tolliday 2012: 17). The level of autonomously developed oil as a share of Japan's total oil imports stood at 11 per cent in 1978 and has hovered between 10 and 15 per cent over the past decade (Drifte 2005, Christoffels 2007: 10, Koike et al. 2008: 1767). Japan's most successful equity arrangement was the AOC, which operated the offshore section of the neutral zones between Kuwait and Saudi Arabia. It produced 300,000 bpd, roughly half of Japan's equity production (Manning 2000). However, in 2000, the AOC lost its concession after it refused to succumb to Saudi demands to increase its investment in the project (Shaoul 2005).

Regardless of failed past attempts, in its 2006 NNES, Japan set a new target for increasing the ratio of self-developed oil in its total imports from 15 per cent to

40 per cent by 2030. Japan's 2010 BEP also targeted an increase in self-developed fossil fuel supply (which also includes natural gas and coal) from 26 per cent to 50 per cent in 2030 (METI 2010a). The NNES and the BEP recognize the importance of increasing the amount of oil developed and imported through Japanese producers in order to improve Japan's energy security. The NNES listed four advantages of pursuing exploration and development activities. First, they contribute toward the long-term stability of the oil supply because of the direct involvement in production and operation. Second, involvement in such activities helps in making timely prediction of changes in the market, since projects are executed under the energy policies of oil-producing countries. Third, it also makes it possible for investors to understand global trends of exploration and development and strengthen business partnerships with others. The last advantage mentioned in the report is the strengthening of a wide-ranging and interdependent relationship between Japan and oil-producing countries (METI 2006a, METI 2010a).

The 2006 target reflects a major structural change in the international energy market owing to various elements concerning both supply and demand conditions. In this context, the target for Japan's self-developed oil is higher than those previously set (Koike et al. 2008: 1773–74). According to Toichi (2006: 17), the 2006 targets for the increase in percentage of equity oil from 15 per cent to 40 per cent by 2030 will be difficult to achieve. Various factors have created impediments toward achieving the target. These include the institutional design of cooperation between government and private industries, as highlighted in Chapter 1; the early history of the upstream industry; the target area of overseas development; and the changing environment in the international oil and gas industry. Japan's situation is even more difficult with the advent of new economic powers, such as China and India and a policy to increase the share of self-developed oil exacerbates global and regional competition for energy. In fact, it remains unclear why the Japanese government places priority on owning the rights to develop oil fields, rather than procuring oil on international markets (Koike 2006).

When it comes to the international market for oil, the cards are stacked against Japan for a number of reasons. Unlike China and other countries such as India, Indonesia, Malaysia or Russia, where oil companies enjoy financial and political support from their governments, Japanese private oil and trading companies do not have the benefit of such extensive government assistance. Thus, they cannot afford to expand their operations overseas. Compared with PetroChina, CNOOC and other state-supported oil companies in the region, Japanese oil companies tend to be small and are distinctly divided between upstream and downstream players. It is to Japan's detriment that its oil industry is unable to enjoy economies of scale that might be gained if its supply and production chains were more integrated (Koike 2006: 46).

Japanese oil companies also face the challenge of dealing with rising costs and difficulty in developing new oil fields. There are not many economically feasible areas left within Japan to explore. Therefore, a company must enter a high-risk venture with low prospects of finding sizeable reserves. The development and production costs of oil fields are increasing because the size of oil fields has diminished. Moreover, the terrain is more difficult for drilling as the newer fields are often located in hard-to-get-at locations such as deep waters or the Arctic. The problem of soaring project costs is an especially serious challenge for Japan, given the country's difficult fiscal situation, thus rendering an additional increase in expenditure on energy development unfeasible (Koike 2006: 46).

Moreover, Japan's security alliance with the United States further limits its policy to increase the share of self-developed oil. The case of Iran's Azadegan oil field illustrates this constraint for Japan. The Azadegan field was discovered in 1999 and is estimated to hold 26 billion barrels of crude oil. Inpex – the Japanese upstream oil company which won the development rights – anticipated that the project would cost $2 billion and produce 260,000 bpd when fully operational, boosting Japan's imports of self-developed oil by 60 per cent. Azadegan's significance grew in 2000, after AOC lost its concession rights in Saudi Arabia. Azadegan offered a way to replace the drilling rights lost in Saudi Arabia as well as contribute to the longstanding goal of boosting the amount of oil Japan imported from its own oil companies. It also provided a response to China's stepped up activity in the Middle East. Expectations for the $2 billion project rose in February 2004, when a more formal agreement was reached with Iran. The agreement gave Inpex – Japan's largest upstream oil company – a 75 per cent stake in the project.

Despite the agreement and backing from Tokyo, the project failed to move forward. One problem was the large number of land mines that covered the area – left over from the 1980–1988 Iran–Iraq war. Inpex claimed that it was unable to commence drilling until the mines were cleared. But the most significant obstacle was opposition from Washington and its growing standoff with Iran. US officials privately and publicly discouraged Japan from moving forward with the project at several junctures since the late 1990s. At the same time, Tehran became increasingly impatient with Japan's foot dragging. Inpex was pressed to commence development or lose its development rights. As milestones were passed without signs of meaningful progress, pressure mounted within Iran to terminate Japan's involvement in producing oil from the field. The head of the Iranian Parliament's Energy Commission announced that Tehran would cancel the $2 billion contract. The issue came to a head in October 2006, with voices in Iran calling for the agreement with Inpex to be scuttled, winning the day. Inpex was forced to relinquish control of the project as its stake was cut to just 10 per cent (Evans 2006: 15). In 2010, Inpex abandoned its stake in the project facing the prospect of being denied access to US financial institutions.

Japan's alliance with the US further affected its relations with Iran since Inpex's Azadegan exit. The aftermath of the Fukushima disaster coincided with the strengthened Western pressure against Iran. Consequently, Japan was strong-armed by the US into reducing oil imports from Iran as a consequence of the strengthened EU and US sanctions regime against Iran. In January 2012, Washington applied pressure on Tokyo to reduce dependency on Iranian oil and

Japan's former Prime Minister Yoshihiko Noda's government has indicated its desire to cooperate (Smith 2012). As a consequence, Japanese oil imports from Iran dropped from 362,000 bpd in 2010 to 314,000 bpd in 2011 and further to 192,000 bpd in 2012 (MoF 2013). While oil imports from Iran accounted for close to 10 per cent of Japan's imports in 2010, by 2012, their share dropped to just over 5 per cent.

Japan had to reconsider working with Iran on the Azadegan project and reduce its oil imports from Iran in light of Tehran's internationally unpopular stance on developing nuclear capabilities. Given Japan's adherence to the philosophy of nuclear non-proliferation and considering Tokyo's important relationship with Washington, it was unable to pursue the opportunity represented by the Azadegan oil field. Japan critically lacks the foreign policy clout of other nations like the US, China, Russia or India. In particular, it is dominated by the US alliance in foreign policy and, unlike China, cannot exploit opportunities for resource diplomacy in countries that are at odds with the United States, such as Iran.

The manner in which Japan seeks to use its oil companies to enhance national energy security resembles the approach taken by Japan to secure overseas equity oil between the 1970s and the 1990s. However, the economic context today is rather different from that of the 1970s, as oil markets have changed greatly. The number of countries and companies that supply oil to international markets has grown dramatically, as has the number of buyers and oil is essentially a fungible commodity, with world oil market behaving as 'one great pool' (Adelman 1984). In this context, the governments of most advanced market economies do not aim to enhance national energy security through the use of oil companies. Rather, the key energy security instruments today include the following: emergency oil stocks, diversified sources of imports, diversified import routes and infrastructure, foreign exchange to meet higher prices, energy efficiency, a diversified fuel mix, diplomacy to keep markets working and naval capacity to keep sea lanes open. Against this backdrop, Atsumi (2007: 30) argues that Japan should refrain from excessive intervention by non-business sectors that could in any way harm market functions or lessen their impact. It is in Japan's best interests to eliminate as far as possible any political dimension from oil in a manner that will reduce its importance as a strategic commodity and strengthen its characteristics as a market commodity.

However, a number of Asian governments, including Japan, continue the practice of promoting overseas investment by their oil companies in the belief that it will provide protection against supply disruptions. Even if overseas investment by Japanese oil companies does not significantly enhance security of energy supply, the Japanese government will likely continue to support Japanese companies as long as the other objectives are also met and oil prices remain relatively high (Andrews-Speed 2012: 30). As a result of the 2011 earthquake and greater need for energy supplies, JOGMEC increased spending by more than $1.12 billion during 2012. This is equivalent to nearly all of the company's upstream investments since its inception in 2004 (EIA 2012b).

Zero-sum Competition Between Asian Oil Importers

The preceding section has argued that Japan's self-developed oil policy has been an expensive and unsuccessful endeavour. However, the Japanese government remains committed to the policy despite historical failures and strong arguments against it. In addition, Japan's *hinomaru* oil policy also encourages competition with China and India, two countries which have implemented similar policies. In the past decade, Japan has been competing with other Asian importers to secure long-term oil supply contracts with suppliers in the Middle East and other regions, often failing to outbid Chinese national oil companies who are backed by deep pockets of their home government.

China and Japan used to have strong energy relations before China transformed into a net oil importer in the early 1990s. The recent years have witnessed an increasingly intensive competition between the two countries over petroleum supplies (Liao 2007). When China became a net oil importer, Sino-Japanese energy cooperation weakened dramatically (Dorian 1995). Although the two governments avoided admitting their rivalry in oil supplies, the dispute over the Russian oil pipelines has clearly indicated the Sino-Japanese competition over oil supplies.

The collapse of the Soviet Union in 1989 was greeted with particular satisfaction in Japan. Not only did it bring about the demise of Japan's most significant military threat at the time, but it opened new possibilities of gaining access to Russia's vast oil and gas reserves. Some of these, such as Sakhalin Island north of Japan, were enticingly close. Other supplies were located further away in Siberia, but nevertheless appeared to offer a way to reduce Japan's heavy dependence on Middle Eastern crude. Throughout the 1990s there were great expectations about tapping Russia's large oil and gas reserves. However, the strategy to diversify energy sources using Russian reserves has been more difficult than expected. The level of competition from China is a factor that Tokyo did not anticipate as it laid plans to tap Russian energy (Goldstein and Kozyrev 2006). As China's energy needs grew in the 1990s, it increasingly looked to Russia as a source of supply. The Chinese government, in conjunction with Chinese oil companies, undertook a determined effort to secure oil pipeline agreements linking the two countries. Much to Tokyo's surprise and consternation, China's resource diplomacy yielded a number of energy supply agreements (Evans 2006: 15).

The bilateral dispute over the East Siberia Pacific Ocean (ESPO) oil pipeline was about two different routes proposed by China and Japan, respectively. One was a 2,240 km line between Angarsk and Daqing signed in July 2001 as an 'interstate agreement', by the Chinese State Planning Committee and the China National Petroleum Corporation (CNPC), on the one hand, and the Russian Energy Minister, Yukos and Transneft, on the other hand. The pipeline involved an estimated cost of US$1.7 billion, supposedly to be completed by 2005 and to provide 400,000–600,000 bpd of oil to China by 2010. The pipeline agreement was further substantiated by the Chinese Premier Zhu Rongji two months later in Moscow, when he and his Russian counterpart Mikhail Kasyanov agreed to start a feasibility study for the project. Not until early 2003 when Japan raised the Angarsk-Nakhodka proposal, the Sino-Russian pipeline agreement was believed to have been completed, as a resolve shown by the two governments 'to consolidate an energy component to their strategic partnership' (Andrews-Speed et al. 2002).

The 3,800 km pipeline from Angarsk to Russia's Pacific port Nakhodka was initially proposed by Russia's state-owned company Transneft, in June 2001, as a means to enter Asian markets, involving an investment of US$5 billion, but no detailed plans followed until Japan showed its interest in the project in late 2002 by offering financial support. Japan's intentions were confirmed by Prime Minister Junichiro Koizumi on his visit to Russia in January 2003, when a six-point 'action plan' was signed calling for cooperation in economics, energy and international diplomacy. Japan's decision to support the Pacific oil pipeline project can be analysed from different aspects. In energy aspect, Japan worried about China becoming a monopoly power to get the Siberian oil and thus lobbied Moscow to allow the oil to be available to the wider Asia–Pacific market. Economically, Japan saw an oil pipeline deal as a breakthrough in Japanese–Russian relations, for the Russian market remained almost untapped by the Japanese. The 'Asian Premium' could also partially explain the offer, as Japan had to pay US$3 billion extra annually for its oil imports from the Middle East. Japan also appeared willing to promote political relations with Russia through energy cooperation, in order to create a favourable environment to settle the bilateral territorial disputes and to confront China's growing political and economic clout in East Asia. Overall, Japan's competition with China over the oil pipelines was not purely based on the concerns of energy security. Instead, it was an unusually aggressive diplomatic offensive driven by both energy-related and non-energy concerns.

After two years of deliberation, the Russian government announced on 30 December 2004 that a pipeline from Taishet, in East Siberia, to Nakhodka, on the Pacific coast, would be constructed by the Russian state-owned company Transneft, with an estimated cost of US$16 billion. Two reasons were believed responsible for the decision. One was that the Pacific route could allow oil being supplied to a number of customers and the other that Japan could provide financial assistance to cover the cost. To placate the disappointed Chinese, the Russian and the Chinese reached an agreement on building a branch pipeline to China, which was confirmed by Premier Wen Jiabao in March 2005. However, this Russian decision did not represent the end to the saga. On his visit to Tokyo in late April 2005, the Russian Minister for Industry and Energy Viktor Khristenko avoided saying which oil pipeline would be built first. This represented a clear example of Russia's tactic to have China and Japan compete against each other in order to reap the greatest financial benefits.

Eventually, Moscow embarked on building the ESPO pipeline without any substantial blueprint of realizing the project's second phase in which the pipeline is extended to the Pacific. Russia's Transneft, backed by the Russian government, is currently building the ESPO, a 4,857 km oil pipeline from Taishet, Siberia to the port of Kozmino on the Pacific Ocean, to export Russian oil to the energy hubs of the Asia–Pacific region. In September 2010, the first section of the pipeline, running from Eastern Siberia to Skovorodino, was completed with a capacity of 600,000 bpd. The 64 km long section from Skovorodino to the Amur River on Russia–China border and the 992 km long section from Russia–China border to Daqing (built by CNPC) were also completed in September 2010. In January 2011, Russia commenced scheduled oil shipments to China, pumping approximately 300,000 bpd. The second section of the pipeline, from Skovorodino to the Pacific Ocean, is scheduled to come online in 2014 and will be able to transport an additional 1 million bpd of crude oil to China, Japan and South Korea (EIA 2012b). In the meantime, oil is transported from Skovorodino to Kozmino by railway. However, there is much uncertainty regarding Russia's ability to be able to meet this capacity, due to insufficient scientific data on the availability of proven reserves in Eastern Siberia, necessary to be validated in advance for commercial production. Moreover, Russia had failed to attract as much Japanese investment as anticipated in carrying forward the ESPO project as a means of countervailing Chinese influence in the development of Eastern Siberia and the Far East (Itoh 2012: 159–60).

States' interests in relative gains make cooperation difficult and international regimes can help the self-interested states to cooperate 'when opportunities for joint gains through cooperation are substantial', because 'states' obsessions with relative gains will diminish' (Keohane 1993). As evident from the ESPO pipeline case, China's transformation to a net oil importer and its growing economic potential turned China and Japan into competitors over oil supplies and also rendered cooperation between the two countries less likely because of their increasing concerns over relative gains. The end of the Cold War not only weakened their shared strategic and political interests, but also posed China as one of the major threats in Japan's security agenda. As a result, China and Japan chose to compete, rather than to cooperate, in securing their petroleum supply. China and Japan have no concrete form of policy adjustment towards energy suppliers at the bilateral level (Itoh 2009: 161).

At minimum, this competition has raised the costs of oil imports for both countries by delaying the development of regional infrastructure, arguably exacerbating both states' energy insecurity. For example, some argue that Tokyo's original impetus for participation in the Azedegan project was to gain a renewed foothold in the Middle East following Chinese gains there; the loss of the stake was seen in zero-sum terms (Shaoul 2005). Instead, Japan's decision to exit the project due to US pressure simply delayed the completion of the project. Likewise, as illustrated above, the competition over the route of ESPO pipeline has developed a soap opera-like quality. The central impact of the ever shifting pipeline routes has been to delay a final plan and to raise costs through expensive stop-gap measures such as China's oil imports by rail and shipment by rail to the Pacific. Finally, the dispute over gas and oil rights in the East China Sea has delayed full production at Chunxiao fields (Manicom 2008). Japan contested China's right to produce resources in the area and on several occasions China has halted development of the fields for diplomatic purposes, which has delayed full production.

The tussle between China and Japan over oil and gas reserves in the East China Sea shows how rigorous competition over energy sources also can result in diplomatic disputes and deepen mutual mistrust between countries. Despite the existence of a compelling common interest in developing the region's energy resources for their mutual benefit, energy concerns have emerged as a significant source of friction in the Sino-Japanese relationship, as both parties have conceptualized energy as a security issue, rather than an economic one and have furthermore apprehended the issue through the lens of an already tense and deteriorating bilateral relationship. While Japan's tendency to conceptualize energy as a national security issue is far from new, its increasingly assertive regional energy diplomacy does constitute a significant departure from Cold War precedents (Phillips 2013: 25).

Overreliance on Middle Eastern Imports

To satisfy its demand for energy, Japan continues to rely heavily on oil imported from the Middle East. Historically, geopolitical uncertainty in the region has underscored the need to diversify import sources away from the Middle East. The underlying motif of Japan's oil import policy has been diversification (Eguchi 1980: 270). Diversification of supplier regions and imported energy commodities moderates transit vulnerabilities, including those that are due to regional turmoil. A more diversified import diversification strategy is one which sources oil from as many suppliers as possible, thereby minimizing the potential for supply disruption. Relying on a single source for oil imports, in general, is far riskier than importing oil from multiple sources. Having multiple suppliers provides security and reduces vulnerability in cases of temporary or permanent disruption of supply (Vivoda 2009: 4616). Should one supplier fall victim to natural disasters, terrorism, war, regime change, or other export damaging events, importers will only experience minor disruptions to their total supply (Leiby 2007).

Since the first oil crisis in 1973, Japan has strived to achieve the best energy mix, through diversification of supply sources, while making efforts to utilize energy more efficiently. The policy sought to reduce dependence on imported oil by diversifying into coal, LNG and nuclear power. It also aimed to reduce dependence on Middle Eastern oil by diversifying into alternative oil import sources. Japan has been successful in reducing its overall dependence on imported oil since the 1970s. Japan's dependency on imported oil was reduced from 78 per cent in 1973 to 40 per cent in 2010, before increasing to 46 per cent in 2012 due to increased demand for oil following the Fukushima disaster (BP 2013). While, prior to Fukushima, Japan has been able to reduce its dependence on imported oil, both in terms of volumes and share of overall energy demand, there has been less success in terms of reducing its reliance on Middle Eastern oil. Since the early 1970s, Japan has attempted to diversify away from the Middle East to provide greater assurances of supply and to protect against OPEC's oligopolistic pricing behaviour. Between 1973 and 1988, Japan was able to reduce the share of imported Middle Eastern oil as a percentage of total imports by from 76.1 per cent to 63.4 per cent (see Figure 3.2). This was possible by increasing the share of non-OPEC oil from Communist nations, Latin America and South-East Asia. Efforts to diversify oil supplies in the aftermath of the oil crises emphasized the Pacific Basin, government-to-government agreements and negotiations with independent producers, such as China and Mexico (Petri 1981: 35). By the mid-1980s, the Japanese have managed to find new or growing sources of supply in the Asia–Pacific region, including China, Australia and Malaysia (McDonald 1986: 148). These countries did not have close political, economic and strategic relations with Middle East nations. For example, China has refused invitations to join OPEC and, in some cases, has adopted pricing behaviour which has undercut official OPEC prices in order to gain a larger share of oil export markets. Although Japan remained heavily dependent on Middle Eastern oil, it was able to spread its oil imports in a way which has provided some, if only marginal, insurances against price increases and supply disruptions (Lesbirel 1988: 299–300).

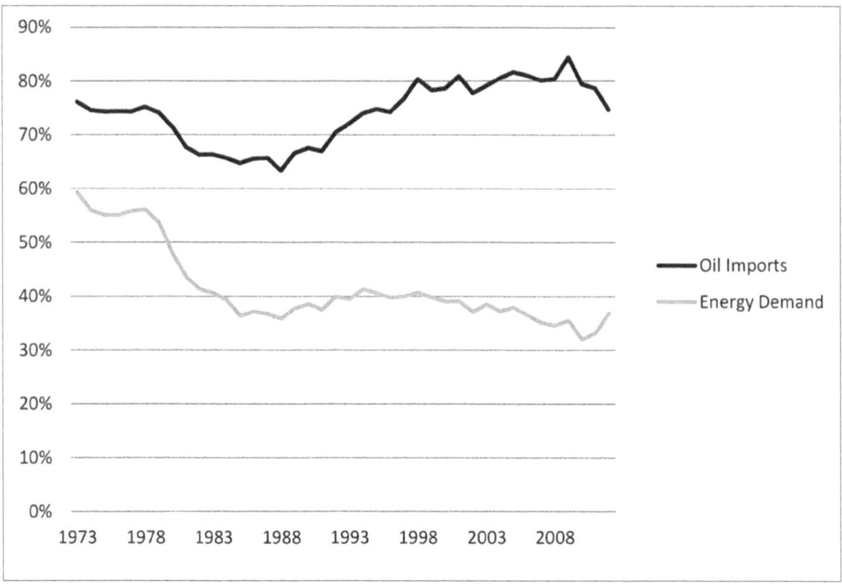

Figure 3.2 Japan's reliance on Middle Eastern oil as share of oil imports and overall energy demand (1973–2012)

Source: BP 2013.

Historically, various political or economic constraints on Japan narrowed the scope for oil import diversification away from the Middle East. In terms of transportation cost, large volume imports from Africa and Latin America have not been feasible because of the long distance to Japan and the limited passage capacity of Panama Canal preventing the scale merit. In addition, during the Cold War, along with other members of the Western bloc, Japan had little access to Communist resource-rich countries, with the exception of China, such as the Soviet Union. As a result, relatively open but not so favourable areas in the Asia–Pacific region remained accessible to Japanese oil companies.

Unsurprisingly, since 1988, Japan increased its dependence on Middle East oil as a proportion of total oil imports. This dependency increased in the 1990s, because non-Middle Eastern oil producing countries such as China and Mexico gradually reduced their crude oil exports in accordance with economic growth in their countries. Consequently, by 1998, Japan's Middle Eastern oil dependence increased to over 80 per cent – a level higher than during the 1970s. Japan's reliance on Middle Eastern oil peaked at 84.5 per cent in 2009, before dropping to 74.7 per cent in 2012 (BP 2013). At the same time, it has to be noted that since the 1970s, Japan has become less reliant on the Middle Eastern oil as a share of its total energy consumption, despite increasing its dependence on Middle East oil imports (see Figure 3.2). The Asia–Pacific has been the only other region with a constant double-digit share in Japan's oil imports. However, due to the region's geological constraints, its share dropped from 20 per cent in 1997 to as low as 8 per cent in 2009. In 2012, the Asia–Pacific region accounted for 15 per cent of Japan's oil imports. However, this figure also includes regionally refined petroleum product imports, much of which originate from the Middle East.

The dramatic rise in global oil prices since 2004 exposed Japan's continued reliance on Middle Eastern oil and sparked a renewed interest in import diversification. In this context, gaining pipeline access to Eastern Siberian oil became more appealing and Japan's oil imports from the Former Soviet Union, mainly from Sakhalin Island, increased from 0.04 per cent to 3.9 per cent of total oil imports between 1999 and 2012 (BP 2013). While Japan's recent forays into North and West Africa and the Russian Far East have had diversification of sources of imported oil as their main aim, Japan has been unable to source more than 7 per cent of oil from these regions combined in any single year (Vivoda and Manicom 2011: 238).

Japan's regional dependence on imported oil remains heavily concentrated on the Middle East. In fact, a higher share of Japan's oil imports originates from the Middle East than for any other industrialized oil-importing country. All of Japan's oil that originates from the Middle East must pass through vulnerable chokepoints. Japan remains dependent on the US for the protection of its oil tankers passing through key sea lines of communication (SLOCs), such as the Strait of Hormuz, the Indian Ocean, the Strait of Malacca and South China Sea (Lam 2009: 119–20). According to Cole (2008: 113), no country is more dependent than Japan on the long SLOCs between the Middle East and Northeast Asia. While it is unrealistic to assume a scenario in which the Middle Eastern countries would stand together against oil exports to Japan even during crises (Itoh 2012: 170), Japan's overreliance on oil supplies from a single region leaves it vulnerable to disruption caused by a potential regional conflict in the Middle East or South China Sea. Japan's particular vulnerability to disruptions in foreign oil supplies will be less of an issue as its overall oil consumption and the oil dependence of the transportation sector decline, but it will not go away completely.

Japan's oil import diversification strategy confronts a number of constraints. First, Japan's geographic position raises the costs of import diversification. Seaborne transportation and refining infrastructure for Middle East crude are well developed, which reinforces Japan's reliance. Japan's state-owned tanker fleet provides a degree of import security by militating against losing oil to alternative buyers and by placing the security of these vessels on Japan's security agenda (Vivoda and Manicom 2011: 241). Japan's seaborne infrastructure could assist imports from South and Central America, although Japan's refineries are currently unable to process Venezuelan heavy crude. Land-based transportation in East Asia remains underdeveloped with ESPO oil pipeline still under construction. Likewise, Japan's overtures to Central Asia contain a high degree of infrastructure spending designed to increase production in order to lower global prices. Direct access to reserves in landlocked Central Asia is impossible. This geographic constraint – the poor availability of oil specific to Japan – creates an incentive for a response that relies on Middle East imports (Vivoda and Manicom 2011: 241–42).

Second, one of the most important constraints of the political dimension of Japanese diversification policy is Japan's relationship with the United States. Japan's alliance with the United States has, on more than one occasion, undermined its efforts to diversify its import sources. Unlike China, Japan is constrained in its ability to import oil from regimes that are at odds with the United States, such as Iran or Sudan. This has not always been the case. As noted earlier, Japan broke with US Middle East policy in the wake of the first oil crisis. Yet, Tokyo is unable to nurture cosy political relationships with oil-exporting states regardless of their political allegiances. If Japan's international oil policy mirrored that of China, it is unlikely that its US ally would be comfortable with Japanese support of pariah regimes such as Sudan and Iran (Vivoda and Manicom 2011: 247).

Finally, Japan simply does not have the capital to compete with China to gain access to new oil-producing regions and, consequently, diversify its oil supplies. In 2011, Japan's total government debt, or the borrowing to cover all past budget deficits, was 230 per cent of GDP, the highest in the Organization for Economic Cooperation and Development (OECD) (Fensom 2012). The mountainous debt reflects years of slow economic growth, many stimulus plans, an aging society, the impact of the global recession and the cost of the Fukushima disaster. Consequently, notwithstanding historical measures adopted to enhance Japan's energy security by reducing dependence on imported oil as a share of total energy requirements, Japan remains fundamentally dependent on imported oil, specifically Middle Eastern oil. Given the major constraints to its diversification policy, Japan's efforts to diversify will likely continue to fail. As a consequence, it will remain extremely important for Japan to maintain and enhance positive relationships with major Middle Eastern oil suppliers, such as Saudi Arabia, UAE and Qatar (Petroleum Association of Japan 2012: 10).

Conclusion

Although Japan's oil demand has dropped since the 1970s oil crises, oil remains the most important energy source for Japan as it is unrivalled in the transportation sector and still heavily used in the petrochemical sector. Consequently, this chapter examined Japan's major challenges related to the supply of and demand for crude oil and petroleum products. Major challenges surveyed in this chapter include continued overreliance on the Middle East, continued inefficiency of Japanese oil companies, dubious viability of Japan's equity (or self-produced) oil policy and increased Asian competition for overseas oil supplies. Unlike in subsequent chapters on other energy sources, much attention has been devoted to the evolution of Japan's oil policy, as Japan's energy security strategy has historically revolved around security of oil supplies. This chapter demonstrates that Japan continues to pursue self-developed, or equity, oil, a historically wasteful strategy and one mired with inefficiencies in Japan's oil companies, with Japan never achieving desired policy goals. The chapter also demonstrates that Japan has been a key participant in the zero-sum competition among Asian oil importers, particularly since the emergence of China as a major oil importer in the early 1990s. These interrelated challenges stem from the Japanese government's treatment of oil as a strategic commodity, which is not surprising given that Japan remains almost exclusively reliant on oil imports and given the country's experience during the 1970s oil crises. Despite Japan's success in reducing oil demand and the share of oil in the overall energy demand, the chapter has also analysed Japan's continued overreliance on the Middle East. Overall, Japan's international oil policy has suffered a number of policy failures despite being a component of Japanese energy security strategy since the oil crises. Equity oil has only ever provided half the oil expected by MITI/METI and Japan continues to be largely dependent on the Middle Eastern oil.

Chapter 4
Natural Gas

Introduction

Natural gas is a fuel with a number of important distinguishing characteristics, which make it attractive to energy importing countries, such as Japan. First, its combustion is relatively clean because it lacks sulphur oxide emissions which accompany the combustion of petroleum and coal. In addition, it is 30 per cent less carbon-intensive than oil and 50 per cent less than coal (Stevens 2010: 4). Given that its carbon emissions coefficient is approximately half that of coal, natural gas has an environmental advantage in the climate change context. Current government carbon-abatement policies and the Japanese government's pledge to lower GHG emissions support natural gas as the cleanest fossil fuel to replace coal-fired capacity. In comparison to oil, with a relatively high reserves-to-production ratio (56 years) natural gas is in ample supply (BP 2013). Finally, the occurrence of gas reservoirs is more evenly distributed on a global scale than the highly concentrated pattern associated with crude oil deposits. From an energy security perspective, natural gas represents a favourable energy source, as it can be procured from neighbouring regions. Consequently, with higher regional reserve spread than in the case of oil, which is primarily concentrated in the Middle East, it is far less politicized. On this basis, natural gas is often considered the energy source that will be the 'bridging fuel' to a sustainable energy system (Kumar et al. 2011). These characteristics have served as grounds for the Japanese government to promote the wider use of natural gas in its energy policy (Sawa 2012: 133).

In particular, the Japanese government has favoured the use of natural gas, or LNG, over other fossil fuels and other sources to replace nuclear energy after the 2011 earthquake. After the Fukushima incident, Japan has replaced much of the idled nuclear capacity with imported LNG. Most of Japan's LNG import infrastructure was not damaged by the earthquake since a majority of these facilities are located in the south and west of the country, away from the earthquake's epicentre. The Shinminato LNG terminal, owned by Sendai Gas, was the only plant closed in March 2011, although the facility was brought back online by December 2011. Therefore, Japan has been able to rely on LNG as a key source of fuel after the accident (EIA 2012b).

The first section of this chapter surveys the evolution of Japan's natural gas policy. Natural gas has played a very important role in Japan's diversification away from oil since the 1970s oil crises and has been a critical fuel source in replacing lost nuclear power in the aftermath of the Fukushima disaster. The following section briefly outlines sector organization, Japan's historical natural gas supply–demand balance and historical LNG import structure. These sections provide background analysis for the remainder of the chapter. Japan's LNG import structure is particularly important in the context of later discussion on LNG pricing. The chapter proceeds by examining major challenges related to Japan's natural gas supplies. The first challenge is for Japan to develop domestic infrastructure in order to keep pace with demand growth. The most pressing challenge for Japan is LNG pricing. Japan's higher natural gas demand for power generation and a tighter LNG global supply market over the past two years have led to a doubling in LNG prices in the Pacific basin, while the prices in North America have dropped. How can Japan challenge the prevailing global LNG pricing arrangements? Are North American LNG imports a 'silver bullet' that will test the long established LNG market orthodoxies in Asia?

Evolution of Policy

Since the 1970s, strict pollution regulations and the uncertainty surrounding nuclear power generation have made natural gas in the form of LNG an attractive energy option for Japan (Morse 1982: 266). Particularly since the oil crises, it has been Japanese policy to promote the process of fuel substitution by increasing the use of natural gas in all sectors, but especially power generation. Natural gas, therefore, has historically been assigned an important role in fuel diversification for energy supply security (Nemetz and Vertinsky 1984: 71). What assisted Japan in the relatively quick emergence as a large LNG consumer and importer during the 1970s and 1980s was that siting of LNG power plants, all of which have been in urban areas, was not subject to the long delays that utilities routinely faced in building hydroelectric, nuclear or coal-fired power plants in rural areas (Gale 1981: 97).

The most successful energy diversification following the oil crises of the 1970s came from the aggressive development of LNG imports. METI provided lavish subsidies to keep imported LNG prices low and to incentivize utilities to use it. As a result, LNG's share in the overall energy demand increased from just over 1 per cent in 1973 to a remarkable 10 per cent in 1990 (BP 2013). The Japanese oil companies did not view this development favourably, because they perceived gas as a way for new entrants to penetrate Japanese energy markets. During the 1990s, increased use of LNG was becoming an increasingly important issue in view of the growing necessity to stabilize GHG emissions; but due to low oil prices, the economic environment surrounding new LNG development projects presented much difficulty (Toichi 1994: 377, Langton 1994: 258). In the 1990s, LNG has also been viewed as a potential source of replacement capacity for deferred nuclear units (Langton 1994: 260).

With the gradual liberalization of the electric power industry in the 1990s, liberalization and restructuring were initiated in the gas industry as well. The first measures aimed at liberalizing Japan's gas market were adopted in June 1994 and went into effect in March 1995. Revisions were made to the *Gas Utilities Industry Law* that allowed gas utilities to sell gas on a retail basis to large-lot volume users, thereby circumventing incumbent gas distribution companies. These contestable customers were defined as those consuming more than two million m^3 per year, representing approximately 720 gas users. The changes permitted non-gas companies (power, steel, oil and other companies with access to gas) to supply large-lot industrial customers for the first time. The revisions also lifted tariff regulations so that parties in the liberalized segment of the market were free to determine price and other contract terms on a negotiated case-by-case basis (Scalise 2004: 165).

Phase two of the Japan's gas liberalization was adopted in May 1999 and went into effect in November 1999. These revisions expanded the contestable market to eligible consumers of at least one million m^3 per year, or approximately 993 large-lot users. Designated general gas companies were also ordered to grant third-party access to gas pipelines on a non-discriminatory basis. The designated companies included Tokyo Gas, Osaka Gas, Toho Gas and Saibu Gas. The smaller gas distribution companies, such as Shizuoka Gas, Hiroshima Gas, Chubu Gas and Hokkaido Gas, have had no obligation to provide access to their infrastructure (Scalise 2004: 165–66). Reforms enacted in 1995 and 1999 helped open the sector to greater competition and a number of new private companies have entered the industry since the reforms (EIA 2012b).

In terms of numbers of new entrants and their market shares, market liberalization has had a greater effect on the gas market than on the electricity market (Miyamoto 2008: 138). However, new entrants to the newly liberalized gas industry have been predominantly incumbent electric power companies. TEPCO, for example, has supplied local city gas companies via its own low pressure pipeline in Chiba Prefecture; Chubu Electric Power Co. has had a joint venture with Iwatani International Corporation and Cosmo Oil to supply LNG via a lorry operation in Mie Prefecture and TEP operates a large natural gas supply business through its subsidiary Tohoku Natural Gas using the Niigata-Sendai pipeline (Scalise 2004: 166).

The 2003 BEP argued for greater use of natural gas, developing natural gas pipeline networks, diversifying supply sources and lowering LNG import prices (Yokobori 2005: 315). However, Japan's NNES of 2006 broke away from earlier support for the substitution of oil by increasing the use of natural gas, the so called 'shift to gas' policy. The Japanese government supported the shift away from oil, as well as the increased usage of natural gas, since the 1970s and virtually until the release of the NNES. The argument behind the shift to gas was that gas is relatively clean, LNG prices are competitive and, by using gas, Japan could reduce its dependence on oil and the Middle East by shifting to LNG from Southeast Asian suppliers. However, the 2006 NNES pointed out that Japan's position in the world LNG market was weakening with growing demand expected in the US, EU and Asia and that the stable supply of gas at stable prices was to come under increasing pressure. METI was signalling that it will no longer push for gas expansion as a 'safe' alternative to oil. The omission of the 'shift to gas' idea was at the time widely interpreted within Japan's energy industry as a political victory for the oil and electric power utilities, who have opposed further expansion of natural gas for some time, vis-à-vis the gas industry. The electric power industry has been reluctant to expand the role of gas within its fuel mix, because of its relatively high price, as well as security of supply concerns and has preferred coal and nuclear energy (Christoffels 2007: 48).

The 2006 NNES and the 2010 BEP policies to reduce the role of natural gas in Japan by 2030 were based on an assumption that the share of nuclear power will increase. Indeed, several analysts have pointed to their linked futures. Toichi (2008) suggests that LNG demand in Japan entirely depends on the status of nuclear power. Miyamoto (2008: 148) argues that the main factors determining natural gas demand in the long-term are the price competitiveness of natural gas against other fuels and the level of nuclear power for power generation sector. Miyamoto et al. (2012: 45) demonstrate that LNG demand in Japan in 2020 will increase from the 2012 level if Japan decides to denuclearize. However, the demand in 2020 is predicted to drop from its peak level in 2012 if Japan restarts its reactors from 2013, decommissions the existing reactors after 40 years of service and only two new advanced reactors commence operation. According to Zhang et al. (2012a: 382), under a scenario in which nuclear power plants shut down after the Fukushima disaster are closed permanently, all nuclear power plants under construction and in planning stage are cancelled and the nuclear power plants in operation are closed at the end of their proposed lifetime, 40 per cent more LNG will need to be imported by 2030 (compared to 2009). On the other hand, based on conservative nuclear power scenario and a scenario in which Japan actively pursues nuclear power, no new LNG power plants will need to be built and LNG demand will drop by 30 per cent by 2030 from 2009 levels.

As per Japan's 2010 BEP, prior to the Fukushima disaster, by 2030, Japan aimed to reduce natural gas consumption by 25 per cent of 2007 levels (METI 2010a). This policy is likely to change to an increased demand for natural gas in the near future in order to replace nuclear power (Hosoe 2012a). In fact, Japan's electricity supply has already switched to gas-fired power plants given their potential to replace nuclear power facilities and reduce GHG emissions (Bhattacharya et al. 2012: 38). It is likely that LNG will play a vital role in Japan's long-term energy supply mix and will be the biggest winner post-Fukushima. Increasing the share of LNG is a quick and politically easy solution. Although there is still a policy hiatus regarding the future role of LNG, stockpiling and supply diversification of LNG have become important policy options in enhancing Japan's energy security following the Fukushima disaster (IEEJ 2012b: 7). One new approach adopted by the Japanese government has been to develop a well-balanced combination of spot, short-term and flexible long-term contracts to import LNG. This idea is backed by the sharp increase in spot and short-term trade during a surplus supply of LNG worldwide.

Sector Organization

In order to enhance supply security in the event of an emergency, the government provides assistance for the construction of pipelines connecting terminals with one another. Given the high cost of LNG storage and the country's geology, along with the difficulties processing boil-off gas, LNG stockpiling is not used as a primary tool to ensure supply security as much as oil. Despite recent government commitment to stockpiling, it is not expected to be used to such an extent in the future and Japan still does not have an official stockpiling policy for LNG. While most of the natural gas is stored in storage tanks linked to LNG receiving terminals, Japan does have five underground storage sites (Yoshizaki et al. 2009). All five are located in Niigata prefecture where various oil and gas fields are located. In an event of an emergency, Japan can mainly rely on inventories – the voluntary stocks of private companies held at LNG terminals, which are equivalent to 20 days of consumption. Japan has 32 operating LNG import terminals, the majority of which is located in the main population centres of Tokyo, Osaka and Nagoya, near major urban and manufacturing hubs and is owned by local power companies, either alone or in partnership with gas companies. These same companies own much of Japan's LNG tanker fleet (EIA 2012b). As its primary means of maintaining supply security, the country takes advantage of diversity of supply sources, contract flexibility and spot and short-term market purchasing.

Since the 1970s, Japanese industries have been actively supporting the development of LNG abroad. The major role has been played by trading houses and power utilities, who have been the largest consumers of LNG (Oshima et al. 1982: 107). One of the unique features of Japan's early LNG imports has been the involvement of large trading houses in contract negotiations, as well as in subsequent transactions. While in the early years an important reason for the involvement of trading houses may have been a lack of experience in international negotiations and development of large-scale international gas projects on the part of Japanese utilities, it can be assumed that their involvement has also facilitated the provision of the necessary funds, both to Japanese utilities as well as to exporters. Here, the well-known interrelations between Japanese trading houses, financial institutions and manufacturers, as well as their close cooperation with the government, may have been major factors contributing to the rapid development of LNG imports (Kiani 1991: 64).

In Japan, domestic rivalries shape three distinct groupings in the world of electric power generation companies. Each of three major electric power companies in Japan – TEPCO, KEPCO and Chubu Electric Company – built relationships with specific trading firms for LNG supply. Historically, TEPCO had long been a recipient of LNG from Brunei and Malaysia through trader Mitsubishi Shoji; KEPCO had special relations with Nissho Iwai through developing Indonesian LNG; and Chubu Electric had special relations with Mitsui Bussan through developing Qatari LNG. At the same time, each of the electric companies did buy gas from 'competing' projects (Hashimoto et al. 2006: 247).

The Japanese gas distribution industry is best characterized as vertically truncated, but within a framework of several regional monopolies. Inpex, Mitsubishi, Mitsui and various other Japanese companies are actively involved in overseas natural gas exploration and production. Japanese retail gas and electric companies are participating directly in overseas upstream LNG projects to assure reliability of supply (EIA 2012b). Japanese regulations permit individual utilities and natural gas distribution companies to sign LNG supply contracts with foreign sources, in addition to directly importing spot cargoes.

The largest LNG supply agreements are held by Tokyo Gas, Osaka Gas, Toho Gas, Chubu Electric and TEPCO, largely in Australia, Southeast Asia and the Middle East. Many of Japan's existing LNG contracts date from the 1970s and 1980s and are set to expire over the next decade forcing Japan to renegotiate term contracts or locate shorter term supply. Some industry analysts suggest that this is driving Japanese firms' interest in acquiring equity stakes in foreign liquefaction projects, in an effort to guarantee future supply (EIA 2012b). The vast majority of LNG supply projects in which Japanese companies have participated have been principally supported by Export–Import Bank of Japan (J-EXIM) and, later through JBIC. In addition, co-financing from J-EXIM/JBIC has played an important role in securing the participation of Japanese commercial banks in financing the projects. Additional government support is also evident in the form of insurance provided by MITI/METI and guarantees provided by JNOC and later JOGMEC, as in the case of oil.

Although electric and gas utilities had formed consortia for traditional LNG procurement, they have recently tended to individually procure LNG under the electricity and gas market deregulation. However, joint procurement by Japanese utilities has been promoted as one of the guiding principles by the government following the Fukushima disaster and joint procurement in North America and Australia has been one of Japan's measures to enhance stable LNG procurement and energy security (Koyama 2012). Following the accident, there were growing indications that Japanese companies were actively seeking fuels needed for the future such as the acquisition of equity interest in LNG projects in Australia, Canada and the United States (*The Economist* 2012a, IEEJ 2012f: 5). The total investment amount for oil and gas acquisitions overseas by Japanese companies from January to April 2012 was ¥652 billion ($8.4 billion), which is almost the total for all of 2011 (*Bloomberg* 2012b).

The downstream activities of regasification and distribution fall within the purview of companies such as Tokyo Gas, Osaka Gas, Toho Gas and Saibu Gas, the four largest companies in terms of volumes, revenues and assets. Each gas supplier provides gas to customers within a franchised service area through its own distribution lines. These companies are vertically integrated with the exception of natural gas production. Osaka Gas, Tokyo Gas and Toho Gas are Japan's largest retail natural gas companies, with a combined share of about 75 per cent of the retail market (EIA 2012b), but there are approximately 240 additional smaller city gas companies throughout Japan. The Japanese gas industry is privately owned but publicly regulated under the *Gas Utilities Industry Law*. Policymaking and regulatory functions are the responsibility of METI.

Supply–demand Balance

Similar to oil, Japan's domestic production of natural gas is very low, amounting to approximately three per cent of the overall demand. Japan's proven reserves of natural gas have almost halved since 2007 (EIA 2012b). Most natural gas fields are located along the western coastline. Total domestic proved natural gas reserves would last for just over two months at current rates of consumption. There is some speculation that Japan may be an early benefactor of research into exploitation of methane hydrates locked into undersea ice, although this will require significantly more research and commercialization of technology before a reliable estimate of reserves and production costs could be established (Nelder 2013). Since 2000, Japan's natural gas demand increased by 4.19 per cent on an annual basis, against the backdrop of an average 0.55 per cent drop in overall annual energy demand during the same period (BP 2013). Regardless of the relatively rapid increase in demand over the past decade, prior to the Fukushima disaster, natural gas accounted for only 17 per cent of Japan's total energy demand (see Figure 4.1), primarily due to the slow pace of development of the gas supply infrastructure in Japan (BP 2013, Miyamoto et al. 2012: 19).

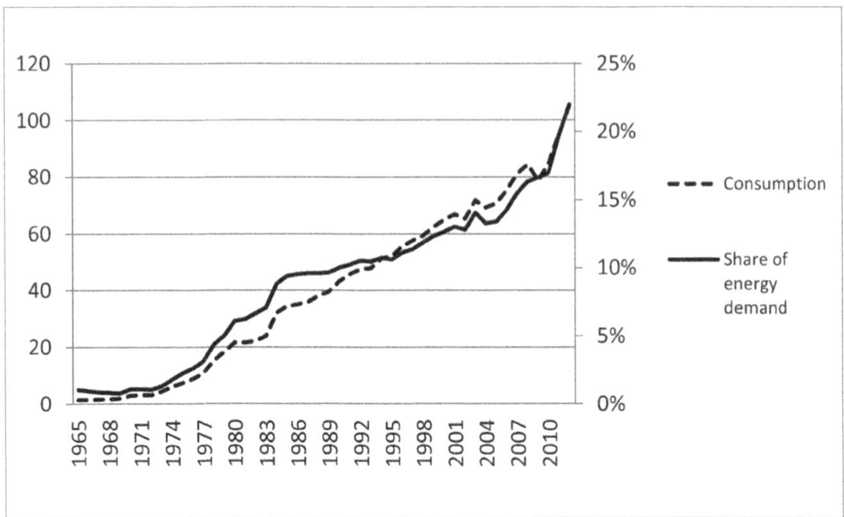

Figure 4.1 Japan's natural gas demand (left axis; mtoe) and share of overall energy demand (right axis; %) (1965–2012)

Source: BP 2013.

In the aftermath of the Fukushima disaster, with the majority of Japan's nuclear power plants idle, LNG became the key fuel for thermal power generation because burning coal and crude oil for power generation was not perceived as a long-term solution given environmental regulations and supply limitations. Consequently, in 2011, Japan's natural gas consumption increased by more than 12 per cent (MoF 2013). Consumption increased despite the fact that Japan's total power demand fell by 4.7 per cent due to post-Fukushima energy conservation and restricted power supplies. In 2011, LNG consumption by power utilities increased by 20 per cent as the delayed resumption of suspended nuclear reactors forced utility companies to increase thermal power generation (Hosoe 2012b: 47, EIA 2012b). In 2012, Japan's natural gas demand increased by an additional 11 per cent on an annual basis. Based on data for the first ten months of 2013, Japan's annual LNG imports are likely to drop by 0.8 per cent, although their value is projected to increase by 15.7 per cent, reflective of higher prices (MoF 2013).

The power sector is the largest consumer of LNG, holding a 66 per cent market share in 2011. City gas demand makes up the remaining 34 per cent of the gas market and consists primarily of industrial, residential and commercial sectors. TEPCO is the largest electric utility and gas importer, holding 44 per cent of the power generation market. Tokyo Gas makes up over a third of the city gas share and is the second largest LNG importer (EIA 2012b). While the future status of nuclear power will be the main determinant of LNG demand in Japan, natural gas demand in the industrial sector will also be one of the key drivers of future demand, as the government encourages oil substitution (Miyamoto et al. 2012: 57). Further liberalization of the electricity market, which could lead to new entrants, a more distributed power generation and a rise in the load factor of self-generators, may also increase LNG demand (Miyamoto et al. 2012: 65).

Import Structure

Japan is the largest LNG importer in the world. In 1980, Japanese LNG imports accounted for approximately 51 per cent of world LNG trade (Morse 1982: 266). By 1985, Japan constituted nearly three-quarters of the world's LNG demand. However, as its incremental take has tapered off and as other countries have entered the market for gas, Japan's relative share has begun to decline. In 2002 its share of the global LNG market was approximately 50 per cent. In 2012, Japan accounted for 53 per cent of Asia's total LNG imports and 37 per cent of global LNG trade (GIIGNL 2013: 8). In 2012, Japanese companies imported LNG from 21 countries and long-term contracts accounted for 78 per cent of Japan's LNG imports. During the past decade, Japan's LNG imports have increased by an average 4.5 per cent annually (BP 2013).

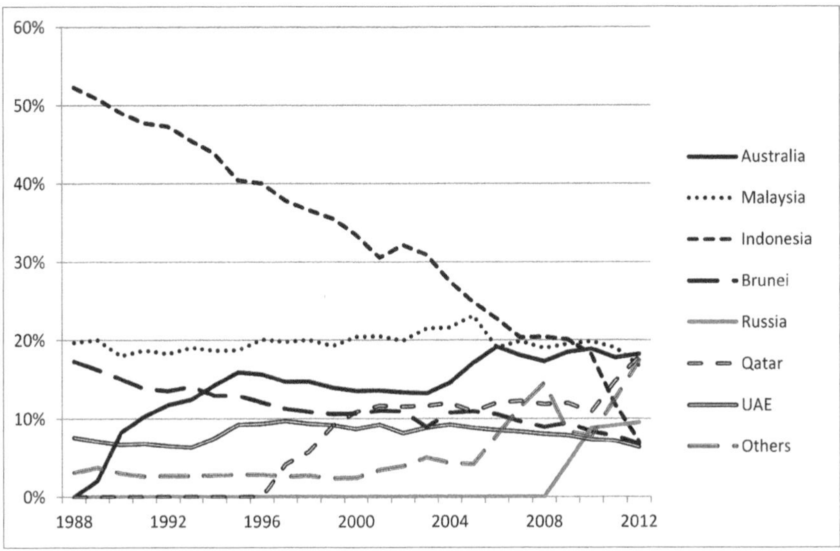

Figure 4.2 Share of Japan's LNG imports by source country (1988–2012)
Source: MoF 2013, BP 2013.

Given that it is still the world's largest LNG importer by some margin, Japan has been the most active in securing new long-term LNG supply contracts. Over the past decade, Japanese companies have signed 56 long-term contracts in six countries (GIIGNL 2013). The Fukushima disaster and its effects on Japan's fuel mix have contributed to changing the landscape of Asia's LNG market. Japan, once considered a mature and saturated market, has become a renewed growing market and will remain the biggest importer of LNG for the foreseeable future (Hosoe 2012b: 45). Japan's regional dependence on imported LNG has historically been heavily concentrated on the Asia–Pacific and specifically on Australia, Brunei, Indonesia and Malaysia (Figure 4.2).

Since Japan has no international pipelines, it is entirely dependent on LNG trade. Techniques to transport and store LNG, developed in Britain in the 1960s, opened the possibility of importing substantial quantities of a clean fuel that previously had to be flared off at the well-head. Japan is a pioneer in the LNG trade, with first imports recorded in 1969 from Alaska. Until the oil crisis of 1973, the prices and distances over which LNG could be imported economically were limited by the relatively low price of crude oil. The escalation in oil prices since October 1973 and the eruption of anti-pollution pressures have greatly enhanced the attractiveness of LNG as a premium, low-polluting fuel for power generation and town gas (Surrey 1974: 214–15). As a consequence, the trade in LNG increased at a rapid pace during the 1970s and Japan has been the dominant actor in the international market. The initial imports of LNG were motivated by the oil crises and a concomitant desire to diversify from Middle East energy supply (Langton 1994: 260). In its drive to develop alternatives to oil in the 1970s and 1980s, Japan played a leading role in launching the global LNG industry.

During the 1970s, LNG was one fuel for which Japan did not have to compete directly with other importers and there was little prospect of a cartel emerging (Gale 1981: 100). During the late 1970s and early 1980s, Japanese electric and gas companies offered long-term contracts for purchases and the Japanese government offered favourable financing via loans and export credits. This government support and its willingness to orchestrate the investment, but also its support for gas infrastructure construction in Japan, were essential for projects to move forward (Hayes and Victor 2006: 325, Jaffe et al. 2006: 472).

The government's interest was based on its desire to diversify energy supplies away from coal and oil. With this support, the first shipments of LNG to Japan came from Alaska in 1969. Later the Japanese government supported projects in Brunei (1972), Abu Dhabi (1977), Indonesia (1977), Malaysia (1983) and Australia (1985). Nearly three-quarters of global LNG shipments were to Japanese buyers in the first half of the 1980s (Hashimoto et al. 2006: 242). In the midst of Iran–Iraq War in 1984 and, despite long negotiations, Japan did not sign a long-term supply contract with Qatar. Although LNG shipments from Abu Dhabi to Japan – over 2 million tons per annum (mtpa) – were not disrupted in 1984 or later in the war, Japanese LNG buyers had other options that did not involve sending expensive $300 million-worth LNG tankers into war zones. In 1985, eight major Japanese gas and electric utilities signed long-term sales and purchase agreements (SPA) with the North West Shelf project in Australia for massive 6 mtpa deliveries, which were later expanded to 7 mtpa (Hashimoto et al. 2006).

By the late 1980s and early 1990s, Japan began to experience regional buyer competition, with South Korea (1987) and Taiwan (1990) becoming LNG importers. The Japanese interest in Qatar was revived by the end of the first Gulf War, with the signing of first SPA in 1992. Japan, with a view to diversifying LNG sources and ensuring a new long-term supply source in the face of new buyer competition for LNG from South Korea and Taiwan, strongly supported the project (Hashimoto et al. 2006). Japan started importing LNG from Qatar in 1997 and Oman in 2001.

Japan's imports of natural gas have not been immune to disruption; in 2001, an important LNG plant in Indonesia, which provided about 30 per cent of Japan's LNG imports at the time, was closed for seven months because of political unrest and separatist conflict in Aceh (IEA 2003: 78, Jain 2007: 31). However, this supply interruption occurred when the field was already in decline and Japan had ample alternative supplies. Still, the perception of Indonesia's instability has caused Japanese buyers to choose Australia and Sakhalin (Russia) for new long-term contracts (Hayes and Victor 2006: 335). Indonesia accounted for about 21 per cent of the world's total LNG exports in 2002 but these have declined since. In March 2006, the Indonesian government announced that it would not renew many of its long-term gas export contracts due to dwindling resources and a preference to have the fuel used domestically. In September 2006, the Indonesian government notified the Japanese companies that it intended to halve LNG exports to Japan by 2010 (Thomson 2009: 29), a threat, which eventually materialized in 2012.

As Southeast Asian countries, such as Indonesia and Malaysia, have reduced their exports to Japan due to the decrease in feed gas for LNG plants and to prioritize domestic supply, exports from the Middle East, Australia, Russia and Africa and re-exports from Europe and the US have grown at a remarkable pace. In 2006 Japan started importing significant volumes of LNG from Egypt, in 2007 from Nigeria and Equatorial Guinea, in 2009 from the Russian Far East and in 2011 from Peru. Japan's LNG import portfolio has become increasingly diversified over the past decade with the number of supplier countries increasing from 8 in 2002 to 21 in 2012. Over the past decade, Japan imported at least 10 per cent of LNG from the minimum of four countries each year. Malaysia, Australia and Qatar supplied over 10 per cent of Japan's LNG imports during the entire period, with Brunei's share dropping from 11 per cent in 2005 to 7 per cent in 2012 and Indonesia's share from 20 per cent in 2009 to only 7 per cent in 2012 (MoF 2013). The drop in Brunei's and Indonesia's shares has largely coincided with an increase in imports from Russia, Qatar, Nigeria and Equatorial Guinea. In 2012, Russian LNG accounted for more than 10 per cent of Japan's imports. Qatar's share increased from 11 per cent in 2010 to 18 per cent in 2012. The share of the United Arab Emirates in Japan's LNG imports stood at between 6 and 9 per cent throughout the past decade, further contributing to Japan's diversified portfolio. Indonesia was Japan's largest LNG supplier from 2002 until 2009 and Malaysia in 2010 and 2011. In 2012, Australia became Japan's largest LNG supplier, closely followed by Qatar (MoF 2013). Between 2010 and 2012, none of the suppliers accounted for over 20 per cent of Japan's LNG imports, which attests to the highly diversified structure of Japan's LNG imports.

At the same time, regardless of enhanced supplier diversification over the past decade (Figure 4.2), Japan remains predominantly reliant on its immediate region for much of its LNG imports. While the regional dependence reached its peak of 91 per cent in the early 1990s, it has been reduced to 58 per cent in 2012, the lowest level since the early 1970s. When Japan first started importing LNG in 1969, a key objective behind this energy policy was to reduce fuel dependency on the Middle East and to diversify supply sources – moving away from the Middle East and increasing other supplies via Asia–Pacific sources – as 85 per cent of Japan's oil imports came from the Middle East. Since then, however, Japan has become highly dependent on Asia–Pacific LNG, which is exacerbated by the fact that most contracts are long-term (over 20 years) for the purpose of supply security (Hosoe 2012b: 50–51).

While Japan may be too dependent on Asia–Pacific LNG imports and on conventional, non-flexible LNG with higher price slopes, Japan has been working actively to reverse this trend. Historically, Japan has been largely reliant on long-term contracts for its LNG imports. However, the share of spot and short-term LNG imports, defined as LNG traded under contracts with a duration of four years or less, has increased from less than 1 per cent in 2002 to 22 per cent in 2012 (GIIGNL 2013: 9). In the 2000s spot and short-term LNG prices were substantially higher than long-term contract prices (Miyamoto 2008: 127, Miyamoto et al. 2009: 27). The average LNG import price exceeds normal long-term contract prices due to the purchase of relatively more expensive spot cargoes. While the effect of this appears to vary according to the conditions of procurement, it may push up the average long-term contract prices by 2 to 3 per cent annually. This appears on the whole to have occurred in the period of escalating oil prices from 2005 to the first half of 2008 (Miyamoto et al. 2009: 26). Following the Fukushima disaster and, until the end of 2011 when the trend reversed, spot and short-term LNG prices were lower than Japanese long-term contract prices, due to the 'global gas glut' created by the expansion of shale gas output in the US and the economic downturn in Europe (Miyamoto et al. 2012: 18).

Spot and short-term LNG procurement by Japan was increased shortly after the Fukushima disaster, with Australia, Russia, Indonesia and Qatar all arranging increasing supplies (Miyamoto et al. 2012: 17). In fact, since the Fukushima disaster, Japan imported LNG under spot and short-term contracts from all but one of its LNG suppliers (Brunei). In 2012, Japan purchased LNG from Algeria, Belgium, Egypt, Equatorial Guinea, Norway, Peru, Spain, Yemen and USA exclusively under spot and short-term contracts (GIIGNL 2013). Since Fukushima, spot and short-term LNG purchases have been driven by the premise that electric and gas utilities operating in regulated markets have an obligation to maintain supplies. For this reason, they are unable to interrupt or cut back supplies to customers when feedstock prices are high. While utilities and gas companies in Japan prefer long-term contracts, from the utilities' point of view, even if the price is one that a utility would normally be reluctant to pay, buying a few cargoes on the spot market is only a small part of the total volume that needs to be procured (Miyamoto 2008: 127). It is under these circumstances that a high volume of spot and short-term transactions for delivery in Asia have materialized since the Fukushima disaster. Although Japan has increased LNG imports in 2012, the country's LNG buyers have shifted focus on their incremental LNG procurement from spot purchases to short- to medium-term contracts, reducing spot market activities and stabilizing spot LNG prices (IEEJ 2012f: 5).

The share of Japan's spot and short-term LNG purchases from most Asia–Pacific suppliers has increased in recent years, mainly associated with debottlenecking exercises and Japan's additional LNG requirements after the Fukushima disaster. Although long-term contracts are not universal in the Asia–Pacific region and small tonnages have been sold under shorter-term contracts during the past decade, given the risks associated with high capital costs in LNG projects it is inevitable that long-term contracts will remain predominant in the future of LNG trade in the region. Since the early 1990s, Japan has come some way towards reducing its dependence on regional LNG suppliers. However, in order to maintain, or further enhance, its LNG import diversification portfolio, the challenge for Japan is to further reduce, or at least maintain, its dependence on its immediate region for

LNG imports in order to challenge the prevailing pricing and contract structures and work towards establishment of the global LNG market, which is discussed later in this chapter.

Domestic Infrastructure Limitations

Although Japan is a large natural gas consumer, it has a relatively limited domestic natural gas pipeline transmission system for a consumer of its size. This is partly due to geographical constraints posed by the country's mountainous terrain, but it is also the result of previous regulations that limited investment in the sector (EIA 2012b). Moreover, Japan's gas distribution system is not well developed in urban areas (Nakatani 2004: 415). Historically, due to the perennial public protest against pipelines, city gas companies have not been able to build the infrastructure needed to distribute gas once it arrives at the large centralized regasification terminals around Tokyo, Osaka, Nagoya and other cities (Gale 1981: 100–101).

Pipeline networks in Japan cover only five per cent of the country's geography and serve one-half of Japanese households. There exists no national trunk line connecting the pipeline networks of the four dominant gas companies, so that, with few exceptions, it is not feasible to purchase or sell gas across regions via pipeline (Soligo 2001: 3). At the same time, new pipeline construction is impeded by the high cost of pipeline construction and the difficulty of securing both construction permits from local and national governments and rights of way (Interview with Akira Miyamoto, 10 January 2013). Companies that wish to build a pipeline in Japan must secure permits from both local and national agencies including, for example, local fire and police departments, METI, the Ministry of Infrastructure and Construction and the Maritime Defence Agency. Firms that plan to build pipelines must also make compensatory payments to landowners to secure rights of way. A respondent from JOGMEC cited 'red tape' in gaining permits to build new gas pipelines as a significant hurdle towards improving Japan's gas infrastructure (Interview with Akira Iishi, 16 January 2013).

Moreover, there are infrastructural limitations on both power-generation capacity and the ability of Japan to import significantly more LNG via existing receiving facilities. Key Japanese importers are already operating at close to capacity and require more investment to meet continued growth in demand. For example, the maximum LNG that TEPCO – Japan's largest LNG consumer – can receive in Tokyo Bay is estimated to be 24 mtpa, which is almost equivalent to LNG imports of China and India combined. KEPCO, Japan's second largest power utility, is also limited in its ability to consume beyond 7.3 mtpa of LNG due to capacity constraints. The utilization rate of Japan's thermal power plants for natural gas is 50 per cent under normal circumstances. In mid-2012, however, the utilization rate stood at 70 per cent. Considering these infrastructural limitations to receiving more LNG, the maximum utilization rate of the power plants that can burn natural gas is estimated at 80 per cent (Hosoe 2012b: 48–49). While some receiving terminals outside Tokyo may have room to import more LNG, they are unable to serve as distribution hubs for Japan's primary demand centres, Tokyo and Osaka. Consequently, Japan should work quickly toward establishing a domestic pipeline network (Tanaka 2012), as rapid development of natural gas infrastructure in Japan will be essential for meeting increased LNG demand (Miyamoto et al. 2012: 65). Substituting gas for electricity in end-use also requires expansion of gas infrastructure that cannot fully substitute for electric power infrastructure.

The Japanese government plans to build the necessary infrastructure to support an additional ten LNG import terminals within a decade (METI 2012b). While there are ambitious plans, the pace at which they are materialized will affect Japan's future LNG demand. In October 2011, Hokkaido Electric Power announced that it would construct its first LNG-fired thermal power station near Sapporo to start operation in 2021. Its LNG receiving facility will be located in Ishikari City, next to the LNG terminal of Hokkaido Gas (Kitagas), which is currently under construction (IEEJ 2011g: 12). Tokyo Gas plans to expand gas supply in the greater Tokyo area by 50 per cent by 2020. The new core facility to develop the required supply network is the Hitachi LNG receiving terminal, scheduled for completion in 2017. It is also developing gas interchange system via a pipeline linkage with other gas distributors to enhance emergency response capability (IEEJ 2012a: 14–15). Moreover, JAPEX has proposed a new LNG receiving terminal to be located in the Port of Soma, Fukushima Prefecture and the terminal will have a ready market in northern Japan. At the same time, the uncertain future of the Japanese power industry makes utilities reluctant to proceed with long-term investment plans to overcome infrastructural constraints. For example, constructing a new 1 GW, combined-cycle, LNG-fired power plant is estimated to cost ¥120 billion ($1.5 billion), which the utilities have not been able to afford since the Fukushima disaster.

LNG Pricing

Even by extractive industry standards, lead times for the LNG projects can be extensive. The high level of capital costs means that long-term LNG contracts are necessary before financing can be negotiated and even then the level of capital costs are such that financing can be tedious and time consuming. The need for delivery of sophisticated LNG tankers is a further complicating factor. A decade can be required from conceptualization through actual first LNG deliveries. The high capital costs and risks associated with extended development times have dissuaded many private sector firms from more extensive equity or debt involvement. For example, massive capital costs involved in LNG production served to discourage more rapid development during the 1970s (Langton 1994: 257). Consequently, there has been a high component of government (or quasi-government) participation in most projects. Export financing of equipment is often arranged by agencies associated with the governments of consuming nations. The large capital costs and the inherent inflexibility associated with this type of system invariably require special contractual arrangements to protect both supplier and purchaser. As a consequence, the international trade in LNG has been characterized by long-term supply contracts of between 15–25 years in length with accompanying take-or-pay clauses. Once in place, these systems offer little opportunity for flexibility in delivery without financial malaise.

Japan mainly imports LNG under long-term contracts with take-or-pay clause. Under take-or-pay, buyers are obliged to take a fixed quantity of LNG over a fixed term regardless of fluctuations in the supply and demand of natural gas. The typical take-or-pay clause states, 'If the total quantity of LNG taken by the buyer is less than the Annual Contract Quantity for the contract year, the buyer shall pay the seller for the quantity deficiency at the average price for each month in that contract year' (Namikawa 2003). Take-or-pay usually accompanies long-term contracts (15–25 years) so as to realize the objective of amortizing the initial investment. The main economic reason why take-or-pay has been used for LNG trade is that it assures a long-term cash flow to amortize the large initial investment of an LNG project and it secures project finance. Take-or-pay suits buyers such as Japanese utility companies that have a delivery obligation to their end users. Furthermore, take-or-pay contributes to energy security policy of a buyer country. It was during the 1970s that a contractual term of 15–25 years and the use of take-or-pay clauses developed into a standard practice and was established as the norm. Take-or-pay obligations have usually been in the range of 3–10 per cent of annual contract volumes and have remained at this level.

Events in the international energy market and practices applied in subsequent LNG contracts have led to LNG prices linked to the mean cost of Japan's crude imports (cif) since the 1970s (Kiani 1991: 66). Crude oil parity became prevalent in the 1980s and provides for LNG prices that are directly linked to cost–insurance–freight (cif) crude oil prices – on an energy equivalent basis (Langton 1994: 262). In 1977, Abu Dhabi and Indonesia commenced LNG exports to Japan. While a fixed price approach was initially employed for the Abu Dhabi project, a price formula that linked LNG to crude oil prices was employed for the 1973 contract with Indonesia. This formula used the average of the export prices (government selling price; GSP) of main Indonesian-produced crude oil as the reference and incorporated an inflation rate of 3 per cent per annum as a variable factor in addition to crude. In 1979, the second oil crisis struck, accelerating the rise in crude oil prices. Coinciding with this, the Abu Dhabi LNG price was changed from a fixed price to one linked to the import prices of crude oil arriving in Japan. At the beginning of the 1980s, OPEC announced a policy of basing the selling price of natural gas on the crude oil equivalent. In response, Abu Dhabi adopted a crude oil price-linked formula that put the LNG price on parity with Murban Crude from January 1980, as a result of which the LNG prices rose sharply. Then in 1983, exports to Japan under the Malaysia 1 and the Indonesian 1981 contracts commenced, both of which employed crude oil price-linked formula. While Malaysia referenced both the average price of crude oil arriving in Japan (Japan Crude Cocktail; JCC) and the official GSP, Indonesia referenced the GSP of Indonesian crude oil (Flower 2008, Miyamoto et al. 2009). Oil indexation for long-term contracts has remained the industry standard in the region because specific contract arrangements have used the price of crude oil, the most widely traded global energy commodity, as a benchmark (Standard & Poor's 2012: 6).

According to Namikawa (2003), take-or-pay contracts indexed to crude oil prices, have fundamental economic drawbacks, especially on the LNG buyer's side. First, take-or-pay with a long contract term reduces the effect of both competition among gas suppliers and cross-fuel competition. This is one of the key reasons why the price of LNG is higher than that of pipeline natural gas where take-or-pay has almost disappeared. Moreover, buyers assume much risk, since they must take a fixed quantity of LNG over a fixed term regardless of the international supply and demand of natural gas as well as the domestic retail demand. Finally, take-or-pay makes it difficult for buyers to find industrial end users since it lacks the flexibility to meet fluctuating demands. In the US and Europe, take-or-pay was once included in contracts to deal pipeline natural gas. However, take-or-pay mostly faded out in the 1980s and 1990s. Nowadays, the market in pipeline natural gas operates according to the law of supply and demand. However, in the case of LNG trade, especially trade to Japan, take-or-pay is still included in most contacts where it plays an important role.

Before the oil crises, the Japanese government was not interested in the clauses of LNG import contracts. Therefore, no policy decisions on take-or-pay in LNG purchase contracts were taken. At the same time, utility companies have thought it best to secure fuel supplies and therefore they have accepted take-or-pay as a measure to guarantee the flow of LNG. Following the first oil crisis in 1973, the rigid, long-term LNG contracts that tied users and producers were accepted as enforcing energy security. The government established a financial support system for Japanese companies to join overseas LNG projects at the post-exploration stage, i.e. concessive loans by the J-EXIM (1974) and surety by JNOC for the loans (1979). Such a system over time could weaken the effect of take-or-pay by increasing the supply of LNG. Nevertheless, the government aimed to increase the domestic demand for LNG at the same time. This shows that the government did not expect the support system would undermine take-or-pay. That is, the government was interested only in a stable supply of LNG (Namikawa 2003: 1331).

In the immediate aftermath of the second oil crisis, Japan was forced to reduce imports of oil. Regarding LNG, the government kept to its policy to increase LNG imports and to secure a stable LNG supply, paying no attention to the attendant problems of take-or-pay. In this period, two systems were introduced to increase the domestic demand for LNG. One was a system of concessive loans by the Development Bank of Japan for LNG importer infrastructure and LNG user plants and the other was a special gas tariff on industrial users. The experience of putting these systems into practice led the government to understand some years later that take-or-pay prevented industries from using LNG. Nevertheless, at this point, the government believed that take-or-pay was necessary to secure LNG. The government had no intention to scrap take-or-pay.

In the 1980s, with ample energy supplies available in the market, the government amended its policy on take-or-pay and began to take effective measures to scrap it (Namikawa 2003). Japan no longer had a sense of urgency about energy security because of the surplus supply of oil which began in 1981 and the low increase in domestic energy demand caused by changes in the industrial structure. The power companies understood that it was an inconvenient fuel for power stations. These companies used LNG only for middle or peak loads because of its higher cost over that of coal or nuclear power during the 1980s. However, the consumption of LNG in middle and peak load periods fluctuates sharply. Since power companies must take a fixed quantity of LNG under take-or-pay, LNG has not been a cost-effective fuel for power stations. At the same time, the government understood that take-or-pay was an obstacle to its own efforts to increase the demand for LNG in the industrial sector. Furthermore, the LNG import contract from Alaska (1969–1984) was about to expire for the first time and preliminary negotiation to renew the contract had started. The Japanese side considered that negotiation as an opportunity to tackle the take-or-pay problems. Additionally, discussions on natural gas pipelines in the US accentuated the problems of take-or-pay.

As a consequence, the government decided to reduce LNG imports and froze construction of LNG-fuelled power plants. This policy produced positive results for Japan in the late 1980s. The first such result was a shorter contract term. The renewed import contract for Alaskan LNG had a term of approximately 5 years starting from January 1985 in contrast to the old term of 15 years beginning in 1969. The agreement to import additional LNG from Indonesia also included a shorter term. A second result was the reform of the buyers' option clause. In the contract for Alaskan LNG, a new buyers' option was agreed upon in 1988 that allowed the buyers the flexibility of taking 6 per cent more or 10 per cent less than the original take-or-pay quantity. In the contract for Indonesian LNG, the rate of buyers' option to take less than the take-or-pay quantity was increased to 7 per cent from the original 5 per cent (Namikawa 2003: 1333).

With an approximately 70 per cent share of worldwide LNG imports (100 per cent in the Asia–Pacific region) in the 1980s, Japan was an influential force in the international supply and demand of LNG. The role of Japan as the only buyer of LNG in the Pacific Rim in the 1970s and the early 1980s offered the opportunity to use virtual monopsony power to break this fuel's price link to oil and/or reduce the effective cost of take-or-pay clauses in LNG sales contracts (Nemetz and Vertinsky 1984). Furthermore, Japanese power companies had potential bargaining leverage, since three-quarters of globally traded LNG was consumed in their thermal power stations during the period. In addition, with many LNG projects looking for buyers, the circumstances of LNG trade were favourable to Japanese buyers. The policy to freeze LNG-fuelled power stations was backed by those circumstances to produce the results mentioned above. The policy, however, did not scrap take-or-pay or challenge crude oil price indexation.

This is because the government failed to pursue any other effective policies. The government remained committed to take-or-pay contracts linked to crude oil prices, as this secured a stable supply of LNG.

If the Japanese government had intervened in each contract by taking advantage of the support system in effect under the circumstances, take-or-pay would have collapsed (Namikawa 2003: 1334). This was particularly the case during the 1990s, as LNG exporters were in favour of a revision of pricing formula that links LNG prices to crude oil prices, citing wild fluctuations in crude oil prices and poor economic viability of new projects that were based on low crude oil prices at the time. Instead, they were pushing for adoption of a sliding-scale system, with price adjustment based on rate of inflation in OECD countries and gas-producing and LNG-importing countries (Toichi 1994: 373). The lack of greenfield projects during the 1990s reflected the belief that prices were too low to justify the necessary multibillion dollar investments (Langton 1994: 263).

Since the 1990s, LNG came into favour again. The government's policy to scrap take-or-pay was toned down. It was predicted that the supply and demand of oil would be tight and that oil prices would increase because of economic development in Asia, the anti-nuclear movement stimulated by the Chernobyl accident and the declining production of oil in the North Sea and in North America. The Gulf War made the government realize the importance of LNG, with reserves more diversified than those of oil. Moreover, it was predicted that the supply and demand of natural gas was becoming tight. Natural gas gained further value because of its low GHG emissions. Finally, Japan was no longer the only LNG importer in the region, since two other Asian countries started to import LNG (Korea in 1987 and Taiwan in 1990). Consequently, the government once again favoured more LNG imports. The policy of freezing the capacity of LNG-fuelled power plants was replaced by a policy of increasing it. The government also considered LNG as an all-round fuel covering the base load to peak load periods and stressed the significance of a reliable and close relationship between suppliers and users. The Japanese government has continued following these broader policy guidelines established in the 1990s.

Historically, Japanese energy policy has played an important role in keeping take-or-pay in LNG import contracts linked to crude oil prices. Take-or-pay raises fuel costs, but utility companies have been able to pass along the cost on the tariff thanks to the market structure in Japan; i.e., a monopolistic utility market with regulated cost-oriented tariffs for utilities. *De facto* monopolistic LNG imports by utility companies, no international pipelines and the low levels of consumption by industry are also elements of that structure. Consequently, utility companies had and, still have, little incentive to oppose take-or-pay clauses. This market structure has played an important role in maintaining take-or-pay.

While long-term sales guarantee remains crucial to the viability of new projects for suppliers, it guarantees secure long-term supplies for buyers, such as Japan. The purpose of take-or-pay, that is, ensuring a stable supply of LNG, has been emphasized in Japan. Japan and other LNG importers have not been able to rely on the market to secure a stable supply of LNG; accordingly, they have depended on take-or-pay arrangements with supplier governments. At the same time, LNG suppliers did not have to depend on the market to find LNG buyers; they have been dependent on take-or-pay contracts, as they provided them with secure long-term customers. Take-or-pay arrangements under long-term contracts have endured for four decades because a global LNG market has not developed. A global LNG market has not developed because long-term contracts with take-or-pay clauses have provided effective and secure terms for LNG trade. In other words, a vicious circle has formed, blocking the growth of a global LNG market and preventing take-or-pay from fading away.

Consequently, satisfied with secure supplies, until recently the Japanese government has not pursued an effective policy to scrap take-or-pay contracts or oil price indexation. However, according to Yokobori (2005: 312), in recent years the Japanese government started considering the contractual take-or-pay clause as inflexible and expensive and started working towards establishing a flexible international LNG market. As a consequence of the Fukushima disaster and Japan's need to replace lost nuclear power with natural gas, Japan's LNG import bill has increased from ¥3.5 trillion in 2010 to ¥6.0 trillion in 2012, as is projected to reach close to ¥7.0 trillion in 2013 (MoF 2013). Due to the unpredictable nature of government policy on nuclear power, the possibility of long-term growth in Japanese LNG demand has led to prices under newly entered long-term contracts being set relatively high by historic standards (Miyamoto et al. 2012: 18). Given that LNG is expected to play a more important role as a primary energy source in the future, a key challenge for the Japanese government is to establish a more economic and flexible supply system.

At first glance, there appears to be an easy solution to the LNG pricing issue for Japan: enhancing diversification of LNG import sources will enhance both the security of supply and go some way towards establishing a more liquid global LNG market and challenging the prevailing contractual arrangements. Given that even small increases in energy prices can yield sizeable economic losses through unemployment and lost income, as well as the loss of value for financial and other assets, diversifying imports to cheaper suppliers is a sound strategy aimed at reducing import costs and building a truly integrated global LNG market. Efficient energy-importing portfolios aim to minimize national exposure to price fluctuations, commensurate with creating optimal overall importing costs (Wu et al. 2009). Yet, such a strategy has not been successful. Vivoda (2014) employed a market concentration measure of diversity to assess LNG import diversification for China, India, Japan, South Korea and Taiwan between 2002 and 2012. Vivoda's findings indicate that five of the major regional LNG importers have improved their diversification portfolios over the past decade. The number of LNG suppliers to regional LNG importers has increased from 8 in 2002 to 21 in 2012. At the same time, Asian LNG prices have tripled. Why have improved LNG import diversification portfolios not resulted in a drop in LNG prices or challenged the prevailing pricing structures?

The most compelling answer can be derived from the existing LNG market structure in Asia. LNG price in Asia is linked to the JCC, or the average Japan crude oil cif import price and Asian importers pay a large premium on LNG prices in other regional markets. The JCC price is the crude oil price referenced by the LNG price formula and is also an indicator of when the LNG price is assumed to be in parity with crude oil prices on the calorific value equivalent basis (Miyamoto et al. 2009: 25). Globally, LNG prices are benchmarked to competing fuels. There have been three distinct and relatively independent markets for LNG, each with its own pricing structure. In the United States, the competing fuel is pipeline natural gas and the benchmark price is either a specified market in long-term contracts or the Henry Hub price for short-term sales. Importers and exporters involved in US LNG transactions are exposed to a significant level of risk given the high degree of price volatility in US natural gas markets. In Europe, LNG prices are related to competing fuel prices, such as low-sulphur residual fuel oil. However, LNG is now starting to be linked to natural gas spot and futures market prices. In Asia, prices are linked to imported crude oil. The pricing formula typically includes a base price indexed to crude oil prices. Currently, regional LNG importers mainly rely on long-term contracts in Australia, Brunei, Indonesia, Qatar and Malaysia, driven by JCC indexation methods. Given that the international LNG market is not integrated, there is a significant price differential among the three major basins (Figure 4.3). Most importantly for Japan and other Asian importers, since 2009, the price in the Pacific basin has been considerably higher than in other two basins. According to data from Japan's Ministry of Finance, the average cost per unit of imported LNG increased by 83 per cent between 2009 and the first ten months of 2013, with increases recorded in every year (MoF 2013).

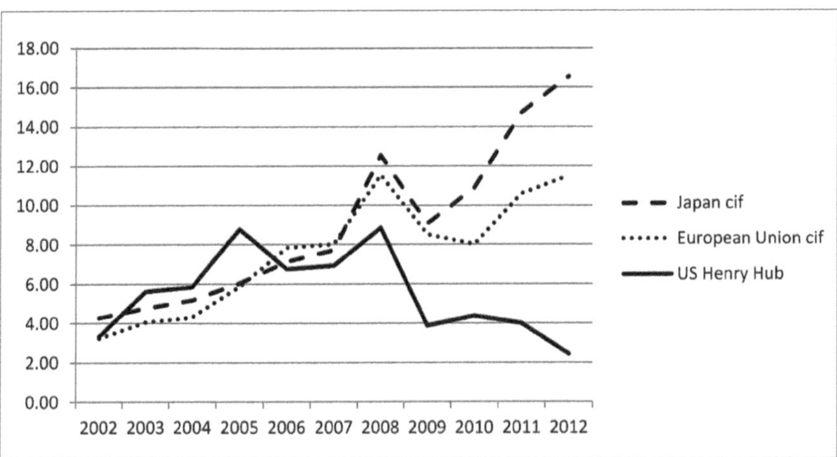

Figure 4.3 LNG price in the three basins (US$ per million Btu; 2002–2012)
Source: BP 2013, GIIGNL 2013.

Given that the competitive relationship between natural gas and oil has weakened over the past four decades, there is recognition among Asian LNG importers that LNG prices should no longer be linked to crude oil import prices. As the LNG market expands, the supply and demand balance of LNG is no longer in line with that of crude oil. It is increasingly important for Japan to revise the oil-indexed pricing system of LNG (IEEJ 2013a: 2). Asian importers and Japan, in particular, have employed two measures in order to challenge the prevailing pricing orthodoxy. The first measure has been an increase in spot and short-term LNG procurement. Much of the improvement in Japan's LNG import diversification and an increase in the number of suppliers have materialized as a consequence of an increase in spot and short-term LNG procurement over the past decade. As mentioned above, in 2012, Japan's spot and short-term LNG imports (defined as LNG traded under contracts with a duration of four years or less) increased from 10 per cent (in 2010) to 22 per cent of the overall LNG imports (GIIGNL 2013); in 2002 and 2006, the share of spot and short-term Japanese LNG imports stood at less than 1 per cent and 8 per cent, respectively.

LNG for spot and short-term trading is available because of excess production (above the fulfilment of the producers' supply obligations) or is available from plants with marginal capacity obtained by debottlenecking and such LNG is often an outcome of the time lag between the production and shipping starts of LNG projects. Spot and short-term LNG also becomes available as a result of conflict between parties to contracts, amortized plants or expired contracts. These factors which free-up spot and short-term LNG are technical or contractual matters as part of individual projects. Spot and short-term LNG trade agreement is an outcome of negotiation and therefore different from the spot trading of oil which led to the establishment of the international oil market. Kiani (1991) argues that unless fundamental changes in technical and economic parameters take place, spot trades of LNG similar to those of crude oil and/or petroleum products cannot develop.

Arguably, according to Namikawa (2003), spot and short-term trade in LNG is unlikely to lead to the formation of a flexible international market. According to *The Economist* (2012b), it will take a lot of spot and short-term LNG to create a liquid global market where the process of physical arbitrage creates a global price across different markets. The reason is that LNG terminals are highly capital intensive projects and are built based on long-term contracts, not on playing in the spot market (Hanser 2012, Wood 2012: 27). While Japan's spot and short-term LNG purchases have increased significantly over the past decade and this has enhanced the overall LNG import diversification, the share of spot and short-term purchases is simply not sufficient to challenge the LNG market orthodoxy in which Asian LNG prices are based on oil indexation.

The second measure employed by Japan and other Asian LNG importers to challenge the traditional pricing and contractual arrangements for LNG procurement, has been to push for LNG imports from the Atlantic basin and North America. Arguably, importing more LNG from North America, will go some way towards the establishment of a fungible global LNG market and will put pressure on oil indexation in Asia. Following the massive shale gas discoveries in North America, Japanese companies have increasingly participated in LNG export projects in the US and Canada. In April 2012, Mitsui and Mitsubishi each signed a Commercial Development Agreement with Cameron LNG to liquefy approximately 4.4 mtpa of natural gas at Cameron LNG terminal in Louisiana. Exports are due to start in late 2016. Also in April 2012, Tokyo Gas and Sumitomo Corporation started negotiations with Maryland's Cove Point LNG project for the procurement of 2.3 mtpa from 2017. In May 2012, Mitsubishi announced a plan to develop an LNG export project to accommodate approximately 12 mtpa from the end of the decade in British Columbia, Canada (named LNG Canada), jointly with Shell, Korea Gas Corporation (KOGAS) and PetroChina. In July 2012, Japanese companies executed binding liquefaction agreements with Freeport LNG, to acquire up to 8 mtpa of LNG from 2017 at a new natural gas liquefaction facility to be built in Freeport, Texas. Japanese companies did not conclude long-term contracts with Sabine Pass LNG, a frontrunner of LNG export projects in the US (KOGAS and India's GAIL did) (IEEJ 2012d: 6; Koyama 2012). Shinzo Abe's visit to the US in February 2013 was partly motivated by Japan's desire to gain permission to import LNG from US mainland (IEEJ 2013c: 10).

Liquefaction tolling arrangements that Japanese companies have negotiated at several LNG export projects in the US are likely to provide a new model of LNG procurement, focussed on gas procurement in North America (based on Henry Hub-linked pricing system) rather than traditional LNG sales into the Asian markets – effectively avoiding discussions of oil-linked LNG pricing (IEEJ 2012f: 5). In fact, LNG from Sabine Pass will be priced in relation to Henry Hub price (IEEJ 2012d: 6). As of early 2013, there are 15 US LNG export projects whose export licensing process in pending, equivalent to a combined liquefaction capacity of 170 mtpa, with 14.7 mtpa thus far destined for Japan (IEEJ 2013a: 9; IEEJ 2013d: 7). The potentially enormous volume of North American LNG exports to Asia could change the LNG pricing system in the Asian market. Increasing North American LNG exports to Asia and enhancing global LNG market liquidity, imply that a new pricing system may be introduced into the Asian market to replace the oil-indexed system (IEEJ 2012g: 4, IEEJ 2012d: 6, Koyama 2012).

Given the discrepancy between Henry Hub gas prices and Japanese LNG import prices as of August 2012, LNG imports from North America could cost 20–30 per cent less than the average cost of those under existing contracts. While market prices fluctuate and the price advantage cannot be guaranteed, the introduction of the new competitive pricing system may contribute to lowering LNG procurement prices under other deals and may be significant trigger for discussions on the pricing system for the existing deals. The additional source of LNG and the new pricing system are likely to be significant factors in improving buyers' bargaining power (Koyama 2012).

Consequently, Japan's new energy policy is expected to encourage the country to import LNG from North America on a large scale in order to improve energy security and diversification of supply sources and pricing formulas. The government believes that Japanese buyers, who have been buying LNG based on oil-indexed prices, should be able to import possibly less expensive LNG based on Henry Hub pricing arrangements. Of particular importance is the fact that volumes from the United States have no destination control, which gives buyers flexibility to manage domestic demand fluctuations. The issue for Japan is that it does not have a free trade agreement (FTA) with the US. In order for US LNG to be a viable option for Japanese importers, the Japanese government has asked the US to take into consideration Japan's unprecedented energy crisis following the Fukushima disaster and allow Japan to import LNG from the US mainland. Currently, the *Natural Gas Act* requires the US Department of Energy (DoE) to evaluate applications to export natural gas and LNG from the United States and then make a determination about whether a permit is consistent with the public interest. This only applies to trade involving countries that have not entered into an FTA with the United States (EIA 2012a). Therefore, Japan is eligible for LNG imports from the United States by either being an FTA member or making the case that exporting LNG to Japan is in the US public interest. It is unlikely that the distinction drawn between FTA and non-FTA countries would survive a challenge under the World Trade Organization. Thus, these differences may slowly disappear by 2020 (Hosoe 2012b: 54–55).

Another important issue to be overcome before Japan and other Asian LNG importers may start receiving LNG from the US mainland, is that the political debate over whether the US should export LNG is resolved in their favour. Some in the US have raised their concern that LNG exports would result in an increase in domestic gas prices. However, the report on the macroeconomic impacts of LNG exports released by NERA for the US DoE in December 2012 concluded that although some income classes and industrial sectors will have to pay higher gas prices, LNG exports will in general bring net benefit to the US economy (NERA Consulting 2012).

While future LNG procurement from North American LNG projects involving Japanese companies are likely to cause a major change in Japan's LNG import mix, matching and rivalling LNG import volumes under the existing contractual commitments from traditional suppliers, such as Australia, is unlikely. With more than 60 per cent of the global planned LNG liquefaction projects located in Australia, with $200 billion worth of largely Asian-financed LNG projects either planned or under construction, there is certainty about the future increase in LNG imports from Australia, mainly under long-term contracts (*Natural Gas Asia* 2011). As much as 70 mtpa of new supply capacity is due to come on stream in Australia between 2014 and 2017, with significant proportion of the production supplied to Japan (IEEJ 2013a: 9). In 2012, Australia became Japan's biggest supplier of LNG, with import volumes predicted to increase significantly over the coming decade. If projects under construction go ahead, Australia will be the largest global LNG exporter by 2017 (Kabede 2012) and the largest LNG producer by 2020, in both instances overtaking Qatar (Bloxham and Hartigan 2012: 1). Since Fukushima, Japanese energy companies have rushed to secure additional supplies of LNG from Australia. If these projects materialize in time and without any significant hurdles, subject to the existing binding contracts, Japan's LNG imports from Australia are likely to more than double by 2020. When future LNG imports from the US are placed in the context of increasing imports from Australia under the existing contractual and pricing arrangements, the possibility for a successful challenge to the existing pricing arrangements in Asian LNG imports seems unlikely during the next decade.

What adds to the challenge is the lack of cooperation among Asian LNG importers. In what is similar to the regional competition for oil supplies (discussed in Chapter 3), regional energy importers also compete for LNG supplies. For example, only Japan and South Korea import LNG from Brunei and significant volumes of LNG from Russia; and only Japan and China import significant volumes of LNG from Australia. The lack of regional cooperation in LNG procurement negotiations with suppliers, endows the latter with much bargaining power to influence the negotiation outcome in their favour, thus preserving the existing contractual and pricing arrangements, which lock-in Japan and other Asian importers into a long-term energy consumption path, allowing limited space for transition to other fuels.

There is much scope for importer cooperation in major LNG exporting countries in the region. Increasing joint development in overseas project and cooperating when negotiating terms and conditions of LNG trade would be beneficial for challenging the existing contractual and pricing arrangements and enhancing energy security cooperation in a region where such cooperation has been missing (Choo 2009, Vivoda 2010). There are three notable examples of such cooperation and there is certainly much scope for further joint investment. First, South Korea's KOGAS and China's CNPC are joint participants in a natural gas development project in Irkutsk, Russia. This project is a large-scale undertaking to connect pipelines from the Kovytinkskye gas field in Northern Irkutsk to supply natural gas to China and Korea. Second, as of September 2012, PetroChina and Mitsui both have a stake in a coal seam gas project in Bowen Basin in Queensland, Australia. Finally, in January 2013, Chubu Electric Power Company concluded a memorandum of understanding with KOGAS for the joint procurement of LNG from Italian oil company ENI. This is the first joint LNG procurement by companies from different nations in Asia. During the contract period from May 2013 to December 2017, both companies will purchase a total of 1.7 million tonnes of LNG, equivalent to 28 tankers (*The Denki Shimbun* 2013). As the LNG market expands, it is increasingly important that Asian LNG importers promote joint procurement and development (IEEJ 2013a: 2). The IEEJ (2013a: 3) pointed out that collaboration between Japan, China and South Korea is urgent in order to de-link LNG from oil prices.

Conclusion

Japan is the pioneer in the LNG trade and remains the world's largest importer. Natural gas has played a very important role in Japan's diversification away from oil since the 1970s oil crises and has been a critical fuel source in replacing lost nuclear power in the aftermath of the Fukushima disaster. In recent years, buoyed by a large regional increase in LNG supplies, LNG prices in the Asia–Pacific region have been considerably higher than in other regions. With increasing demand for LNG in Japan since the Fukushima disaster, Japan has been faced with increased costs for imported LNG. This chapter evaluated Japan's major challenges associated with natural gas supply and demand. The major issues discussed in this chapter include the adequacy of domestic natural gas infrastructure in order to meet demand growth, LNG pricing and continued reliance on oil indexation and the lack of regional cooperation. The key question posited in this chapter was whether and how can Japan break away from the established LNG pricing formula in the Asia–Pacific region. The analyses demonstrate that LNG imports from North America may go some way towards challenging the prevailing LNG market orthodoxy in the region. However, the change will require significant volumes of North American LNG to flow to Asia, which is unlikely to materialize before the end of this decade.

Chapter 5
Coal

Introduction

Historically, the main attraction of coal has been its plentiful supply in the Asia–Pacific region and close proximity of coal-rich countries to Japan: Australia, Indonesia and China (Eguchi 1980: 266). In fact, in 2012 the Asia–Pacific region accounted for 68 per cent of the world's total coal output (BP 2013). Coal has the major advantages of stability and low cost of supply because of its abundance. At the same time, burning coal for electricity generation is emissions-intensive and the challenge for Japan and other countries that are heavily reliant on coal, is to either reduce demand for coal or to aggressively push for clean coal technologies, which would result in lower emissions associated with its use. With Japan's commitment to GHG emissions abatement since the Kyoto Protocol was signed in 1997, the government has been committed to reducing Japan's reliance on coal, the most emissions-intensive fossil fuel.

Yet, coal remains the mainstay in Japan's energy supply mix, maintaining its status as the second largest energy source. The fuel has been particularly useful in the aftermath of the Fukushima disaster in order to replace lost nuclear power and reduce costs of LNG imports. Australian thermal coal has been crucial in this context, with Japan relying on Australia for 70 per cent of its imports. Consequently, this chapter examines two major challenges related to Japan's coal use. First, if coal is to continue to play an important role in Japan's future energy mix, how can its use become less emissions-intensive? In other words, is there potential for new breakthrough technologies to deliver the highly anticipated reduction in emissions from coal-fired power plants? Japan is already the world leader in R&D into these technologies and it is anticipated that it will continue to devote significant resources into development of clean coal technology. In their survey on energy security perceptions in Asia, Sovacool and Vivoda (2012: 958), found that R&D on new and innovative energy technologies was perceived as the most highly rated dimension of energy security in Japan. Their findings highlight the perceived need to further increase spending on R&D. Second, are there any dangers associated with Japan's continued overreliance on Australian coal and can they be overcome? Before turning to these challenges, the chapter begins with a brief survey of Japan's coal policy and analysis of industry organization and coal supply–demand structure.

Evolution of Policy

While it was repeatedly proved that Japan had no significant reserves of oil and natural gas, the country had moderately proportioned domestic reserves of low-grade coal. These reserves were, however, of no use whatsoever to the heavy and chemical industries that dominated early post-war reconstruction, for all their attention was fixed on high-grade metallurgical/coking coal, an essential input for top quality steel. During the 1950s, the Japanese classified oil as a supporting industry, which was understandable since Japan was still mainly coal-powered (Dargin and Lim 2012: 117). However, the post-war policy emphasis upon indigenous coal resources encountered resistance from the expanding industrial sector. Government intervention during 1955–1961, aimed at protecting the domestic producers of coal by restricting petroleum imports, did not withstand the competitive pressures of inexpensive and environmentally preferred foreign oil. As Japan's industrial economy slowly opened to foreign oil, the public treasury bore the escalating costs of the state acting as a guarantor. The political benefits of providing unlimited state subsidies to coal producers seemed always to outweigh the staggering economic costs (Samuels 1987: 134). The increasing price competitiveness of imported oil compared to local coal, led the MITI to reduce its support for the domestic coal industry.

With prices of imported oil and coal falling, the end of Japan's domestic coal industry commenced with labour disputes in the coal mines during the late 1950s (Leaver 2009: 123–24). As a consequence, the domestic coal industry entered a period of rationalization through the orderly closing of uneconomical mines (Nemetz et al. 1984/1985: 555). Domestically-mined coal gave way to inexpensive and abundant supplies of imported oil following import liberalization in 1961 (Culter 1999). Japan's domestic coal industry struggled to survive through the long boom in the face of increasingly free competition from imports of both oil and coking coal. Coal usage in Japan expanded until 1967, when environmental regulations, the high cost of domestic coal production and the ample supply of inexpensive oil dampened prospects for major increases in coal use (Gale 1981). In 1968 a decision was finally made to gradually phase out Japan's inefficient coal industry in response to the increased costs of domestic coal. At the same time, the government enforced a system of protection for the coal industry, which enabled it to sell its coal well above the international price and required major coal consumers to purchase a designated proportion of their coal demand from local mines (Perkins 1994: 596, Surrey 1974: 209–11).

While for many years, coal formed the backbone of Japan's economic development, the dangers and costs of mining became increasingly expensive for the industry and government. In 1973, coal was not given high priority in the alternative energy mix because of a lack of cooperation among coal producers, environmental groups and experts knowledgeable about pollution technology (Morse 1982: 267–68). Following the oil crises of the 1970s, increased coal imports were necessary to supply the additional coal-fired power plants under construction or planned and, to a lesser degree, for other industrial uses (Gale 1981). There was a marked shift from oil to coal in the supply of energy after the second oil crisis in 1979, resulting in higher imports and consumption. To increase coal use during the 1970s, the government relied on two policies. A coal import policy was developed, promoting direct Japanese participation in overseas coal development and long-term contracts for imports. The purpose of the policy was to guarantee a stable supply of coal for as far into the future as possible. The second policy line on coal was the promotion of coal use in the industrial and power generation sectors. The government offered subsidies to partially compensate for the cost differential between coal and oil and it provided preferential loans for the construction of newly planned coal-fired stations (Eguchi 1980: 268). By the late 1970s, Japan was the world's largest coal importer. Virtually all imports were metallurgical/coking-grade coal used by the steel industry (Gale 1981: 104). Only two utilities – Kyushu and Hokkaido – burned coal, relying almost exclusively on domestic coal mined on those two islands (Gale 1981: 102).

As a consequence of the first policy line, Japanese companies increased their investment in foreign coal mines and by the early 1980s the supply from such mines accounted for one third of total coal imports (Oshima et al. 1982). The major role was played by utilities and iron and steel companies, who were the largest consumers of coal (Oshima et al. 1982: 107). After the oil crises, electric utilities were convinced that nuclear power was cheaper than coal and were opposed to a return to coal (Gale 1981: 103). However, they were eventually pushed into the coal importing business by MITI. Only when MITI hinted that it would help to finance the acquisition of large reserves of coal in Australia and encourage government's minority shareholdings in a number of companies did the utilities band together in 1979, through the Federation of Electric Power Companies of Japan (FEPC), to establish the Japan Coal Development Company, a coordinating body tasked with securing long-term contracts for coal for the utilities (Gale 1981: 103–4). As a consequence, the alacrity with which Japanese utilities and other companies were pursuing LNG during and after the oil crises was not matched in intensity by their quest for coal (Gale 1981: 101). As coal-fired thermal plants were encouraged by the government, large investments were required to develop overseas coal production, build an infrastructure for the drastic increase in coal imports and covert oil-burning industrial facilities to coal (Nye 1981: 217). Since coal-fired plants could not be easily built near major population centres, expensive transmission lines had to be built (Gale 1981: 103).

It was only in 1986 with global changes in coal production and exchange that Japan finally decided to shut down nearly all domestic coal mines in favour of coal imports (Culter 1999). This was despite changing market forces that clearly warranted the termination of the industry (Lesbirel 1991). Economic security and regional employment concerns were used to justify the continuation of price differentials in the coal industry. These justifications were not economically objective but were ideologically motivated and politically determined. The power industry, which has been one of the most sensitive to resource security concerns, saw little economic security value in continuing consumption of domestic coal. The ideology of accepting higher prices in order to enhance economic security and prevent unemployment, while having some merits in terms of energy diversification, was seen as very costly and inequitable in the case of coal (Lesbirel 1991). Most domestic coal production in Japan ceased in 2001 (IEA 2008: 99). The slow death of Japan's domestic coal industry highlights the protracted nature of energy transitions.

Since the cost of coal has been much lower than LNG, the share of coal-fired power has increased from 14 per cent in 1998 to 24 per cent in 2006. However, the Japanese government has been reluctant to increase reliance on coal-fired power because of Japan's Kyoto Protocol GHG reduction targets, which was adopted in 1997 and came into effect in 2005. In the late 2000s, when several Independent Power Producers (IPPs) attempted to build new coal-fired power plants, they could not get the government approval due to environmental reasons. Environmental assessments have long presented a hurdle for businesses hoping to construct coal-fired plants. Orix and Toshiba were forced to re-examine and scrap their coal-fired thermal plant projects in 2006 and Nippon Kasei Chemical in 2010, as burning coal was deemed to produce too much CO_2. In fact, no new coal-fired thermal plant construction project has been approved in the decade preceding the Fukushima disaster (*The Yomiuri Shimbun* 2013). Moreover, in November 2002, the Japanese government announced that it would introduce a tax on coal for the first time, alongside those on oil, gas and liquid petroleum gas (LPG) in METI's special energy account, to give a total net tax increase of some ¥10 billion from October 2003. While the taxes in the special energy account were originally designed to improve Japan's energy supply mix, the change was part of the first phase of addressing Kyoto goals by reducing greenhouse gas emissions. More broadly, government coal policy has been focussed on support for clean coal technologies and upstream coal resource development in other countries (IEA 2008: 100).

The 2010 BEP recognized that Japan would have to continue relying to a substantial extent on coal. However the government decided to take several steps to reduce CO_2 emissions from coal. It would promote the commercialization of new, more efficient coal burning technologies, such as integrated gasification combined cycle (IGCC) and require that all new coal plants achieve emissions levels comparable to IGCC. It would also accelerate the development and commercialization of technology for carbon capture and storage (CCS) technologies and require that new coal plants be CCS-ready and then be equipped with CCS technology as soon as it became available (METI 2010a). Under Japan's 2010 BEP, the 2030 target for coal was reduced to approximately 10 per cent of electricity generation. To reduce greenhouse gas emissions in the industrial sector, the government would promote the substitution of natural gas for coal and petroleum. Assuming no increase in steel production, which is a major source of Japan's CO_2 emissions, the BEP anticipated that the industrial sector could achieve a 25 per cent reduction in CO_2 emissions by 2030 (METI 2010a).

Prior to the Fukushima disaster, the government placed priority on reducing the environmental impact of coal burning by promoting the development of clean coal technology. Following the Fukushima disaster and recognizing the need for additional fossil fuel-generated electricity to replace lost nuclear power and increasingly expensive LNG, the Japanese government considered policy reforms in order to make coal-fired power plants easier to build and expand by relaxing its procedures for assessing their environmental impact. No new project has cleared the government's environmental impact assessments since chemical manufacturer Tokuyama Corp. won approval in 2009 to expand a coal-fired plant. TEPCO plans to construct coal-fired plants as an alternative to nuclear plants and is expected to file for approval during 2013 under new assessment procedures (*Platts* 2012). Following the Fukushima disaster, the government was divided on the opinion on whether to promote the plan to allow for new coal-fired power plants. The METI was in favour of the plan, while Environment Minister Nobuteru Ishihara was against it (*The Yomiuri Shimbun* 2013). However, by early 2013, it became apparent that Japan's government has been encouraging utilities to burn more coal by relaxing environmental rules and to build power plants that use technology to limit emissions. In late April 2013, the government eventually relaxed environmental rules on building new coal-fired power stations. With weaker yen and coal prices estimated to be 10 per cent below LNG prices per energy unit, according to METI Minister Motegi, 'This policy is important to diversify Japan's energy options while lowering fuel procurement costs' (cited in Tsukimori and Maeda 2013).

Industry Organization and Supply–demand Structure

Similar to oil and natural gas, Japan depends on imports for more than 99 per cent of its domestic coal needs. Japan is the second largest coal importer in the world after China. Coal accounts for approximately 25 per cent of electricity generation. Japan has 16 thermal coal power plants with a capacity of over 1 GW, in addition to two mixed coal/fuel oil power plants (FEPC 2012: 16). While most of the coal-fired fleet of power stations in Japan is relatively modern (1990s or later) there are a number of older generating units (including a couple of coal/oil units from the 1970s) and some from the early 1980s. That means that a number of older plants will need to be retired in near future. TEPCO will add 2.6 GW a year of coal-fired power from two new plants and electricity bought from two units owned by TEP that restarted after being damaged in the earthquake. The No. 2 unit at TEPCO's Hitachinaka plant, which can produce 1 GW and the 600 megawatt (MW) No. 6 unit at its Hirono facility began operating in April 2013 (*Bloomberg* 2013). The Electric Power Development Company (J-POWER) has several plans for new coal-fired generating stations.

Japan's demand for metallurgical/coking coal used for steel production has been stagnant since the early 1970s, while demand for steaming/thermal coal has increased exponentially during this period. While metallurgical coal accounted for approximately 90 per cent of Japan's coal demand during the 1970s, thermal coal currently accounts for approximately 70–80 per cent of Japan's coal demand/ imports. The Fukushima disaster resulted in a near shutdown in Japan's nuclear industry and supported demand for other sources of energy, including thermal coal. Decreased electricity demand due to restricted industrial usage and power shortages have seen coal shipments halted through *force majeur* declarations or diverted to other power stations. The earthquake caused damage to a number of coal-fired plants, while the tsunami damaged coal unloading terminals located on Japan's eastern seaboard (EIA 2012b).

As a consequence and despite early predictions that Japan's coal demand will increase in the immediate term after the disaster (Behrmann 2011), Japan's coal demand has dropped by 5 per cent in 2011 (MoF 2013). This decline in demand in 2011 (see Figure 5.1) applies mainly to coking/ metallurgical coal as Japan's industrial production and manufacturing sector declined following the disaster. The amount of coal consumed by TEP and TEPCO decreased by 3.78 million tons, while the amount of coal consumed by J-POWER and other utilities increased by 1.57 million tons. Consumption by TEP and TEPCO fell due to the earthquake related stoppages in import terminals and power stations along the Pacific coast (Sagawa 2012: 2). However, in 2012, Japan imported 185 million tons of coal, a 10 million ton increase from 2011, including record levels of thermal coal (MoF 2013).

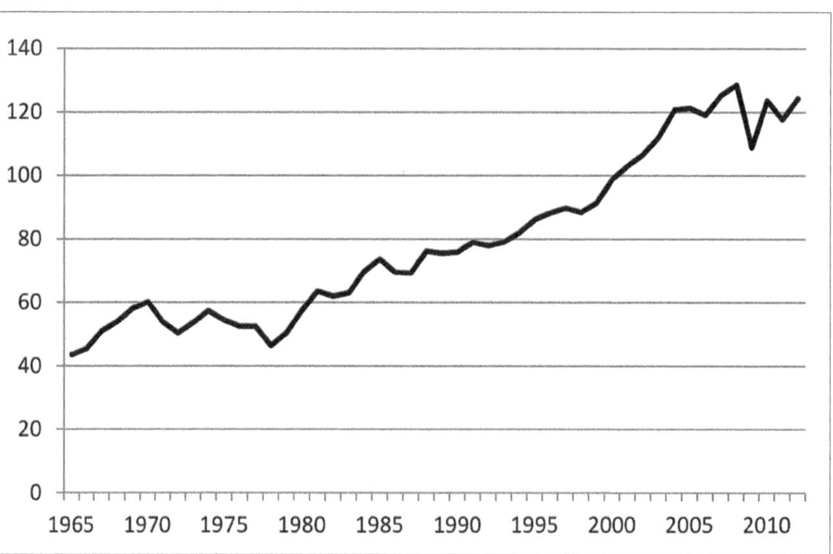

Figure 5.1 Japan's coal demand (1965–2012; mtoe)
Source: BP 2013.

As part of its efforts to secure stable coal supply, Japan conducts government-to-government policy dialogue with coal-producing countries such as Australia, Indonesia and Vietnam and helps the transfer of clean coal technology to coal-producing countries in Asia, such as China, Indonesia and Vietnam (IEA 2008: 87). Australia, in particular, provides the majority of Japan's coal imports. As of late 2012, 14 Japanese steel mills and trading houses have had stakes in 56 operational coal mines in Australia, nine of which are under expansion and in seven new projects.[1] This is a significant presence given that Australia has approximately 107 privately owned coal mines (EIA 2011b). Japanese steel mills and trading houses have established their presence in Australia following the transformation in Japan's status from a coal producer to the world's largest coal importer during the 1970s (Lesbirel 1991).

In fact, Japanese steel mills and trading houses are the world's largest coal importers. They were able to use their combined power to dictate the terms of coal market exchanges with buyers during the 1980 and 1990s. In the past, this buyers' cartel had engineered an oversupplied market characterized by constantly falling prices (Bowden 2012). By 2000, however, this strategy had become counter-productive, as low prices fostered the emergence of a powerful Australian-based selling oligopoly that also controlled thermal coal supply in Indonesia, Japan's second largest coal supplier (Bowden 2012, Bowden and Insch 2013). Historically, the power utilities have relied on steel mills and trading houses to negotiate the annual price for coal before finalizing their own purchasing arrangements. For more than two decades, the Japanese power utilities had benefited from the steel mill and trading houses' buying strategies, which enabled them to set thermal prices at a discount to the annual coking coal benchmark. In 2001, the power utilities' strategy of 'free riding' ended when sellers refused to be bound by the changes in the coking coal rate, forcing a 20 per cent price increase (Bowden 2012). Nevertheless, settlements between Japanese power companies and major producers in Australia remain important in the thermal/steam coal market. Meanwhile, negotiations between BHP Billiton and Japanese steel mills (Nippon Steel and JFE) are still critical in the metallurgical/coking market.

Coal Imports: Overreliance on Australia and Increased Competition from China and India

Historically, the dependence on the Asia–Pacific and particularly Australia has dominated Japan's total coal imports more than the regional dependence on any

1 These companies are Idemitsu Kosan (3 operational projects and 1 new project), Itochu (8+1), Japan Coal Development Company or JCD Australia (2), J–POWER or Electric Power Development Corporation (5), JFE Holdings though JFE Shoji Trading, JFE Steel and JFE Minerals (8+1), Kokan Kogyo (1), Marubeni (8+1), Mitsubishi (9+3), Mitsui Coal (18), Nippon Steel Trading or Nittetsu Shoji (6), Shinsho Corporation (2), Sojitz Corporation (6+2), Sumitomo or Sumisho Coal Australia (5+1), and Tokyo Boeki (1).

other energy fuel import. The share from the Asia–Pacific declined from 94 per cent in 1975 to 82 per cent in 1985 and then increased to 96 per cent by 1999. In 2012, the dependence on the Asia–Pacific region stood at 91 per cent (MoF 2013). Within the Asia–Pacific, changing import shares revolved around Australia and East Asia. In contrast to other energy imports, the diversification of coal imports has revolved around changing dependencies within the Asia–Pacific (Lesbirel 2004: 6).

While there has been a trend in recent years of increased imports from Russia and the United States, the bulk of imports are still sourced from Australia and Indonesia. In fact, over the past decade, Australia accounted for approximately 60 per cent of Japan's coal imports (Figure 5.2). Indonesia's share stands at around 20 per cent. Japan's overreliance on Australia for its coal imports is not a new phenomenon. In 1980, 63 per cent of the imported thermal/steam coal and 41 per cent of the imported metallurgical/coking coal came from Australia (Morse 1982: 268). However, with China's transformation from a net coal exporter to the world's largest coal importer over the past decade, Japan's coal imports from China have dropped significantly, leading to continued overreliance on Australia. In 2012, 71 per cent of Japan's thermal coal and 53 per cent of coking coal imports originated from Australia (MoF 2013). While coal exports from Indonesia are expected to plateau due to increased domestic demand, exports from Russia, Mongolia, Canada and the United States are expected to increase over the coming years. However, Australia will continue to play a central role in the supply of coal to Japan and the rest of the Asian market (Sagawa 2012: 6).

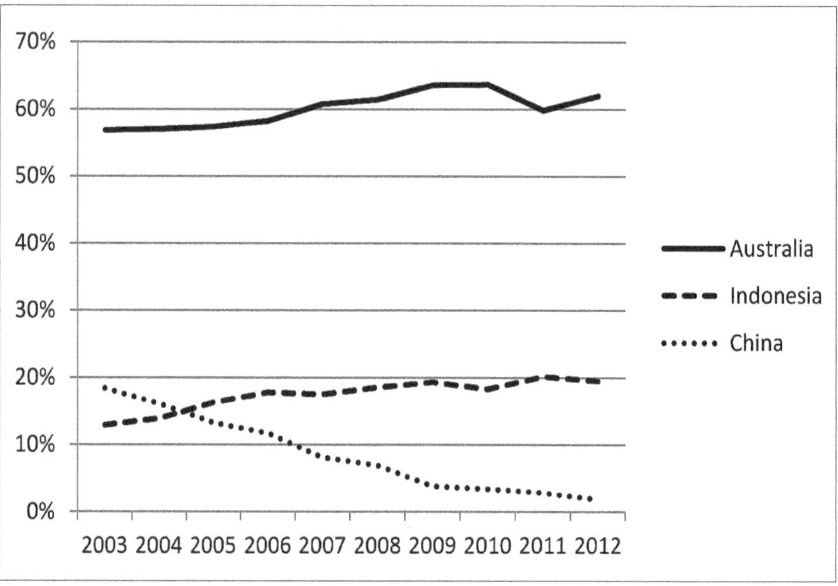

Figure 5.2 Japan's coal import dependence on selected suppliers (2003–2012)
Source: MoF 2013.

Historically, Japanese trading companies used government investment concessions and subsidies to help create oversupply in the seaborne coal market, with buyer domination prevalent until the turn of the century (West 2013). The long era of buyer domination ended abruptly in 2001 when Canadian and Australian producers insisted on a 7.5 per cent price increase. The tipping point in this shift in the balance of power had been brought about primarily by the withdrawal of Canadian supply following the closure of mines (Bowden and Insch 2013: 82). By the late 1990s only a small number of super-efficient mining conglomerates proved capable of surviving in the low price environment that the Japanese steel mills and trading houses had created. For the Japanese steel mills and other Asian coal companies the emergence of a seller's market dominated by an Australian oligopoly happened at an unfortunate time. China, a net exporter of coal during the 1990s, became a net importer from 2006–2007.

By the new millennium, the large Australian producers, notably BHP Billiton, Rio Tinto and Xstrata (formerly Glencore), effectively dominated the Pacific coal trade (Bowden and Insch 2013). Currently the international coal market is highly concentrated. The 'Big Four' (Anglo American, BHP Billiton, Rio Tinto and Xstrata) account for more than 50 per cent of the international thermal/steam coal market, while BHP Billiton alone has a 30 per cent share in the international metallurgical/coking coal market (Energy Charter Secretariat 2010: 25). Asian buyers, including Japan, remain heavily dependent on the Australian oligopoly for supply in the coal trade (Bowden 2012: 20). Such is the importance of Australia that long-term contract prices in the Asia–Pacific are benchmarked against steam coal spot prices in Newcastle – freight on board (FOB) prices of representative Australian coal. Introduced in 2002, the Newcastle index is based on FOB steam coal prices at the Newcastle terminal in Australia, aiming at establishing itself as the benchmark for the Asia–Pacific steam coal market (Energy Charter Secretariat 2010: 27). This dependence is likely to stay, given that the composition of Japan's coal imports centred on Australia and Indonesia, where Australian companies also hold major stakes in various coal projects, will remain unchanged moving forward (Sagawa 2012: 3). Consequently, for Japan, it is essential to strengthen the relationship with main coal supply countries, such as Australia and Indonesia (Ando 2012).

Moreover, the emergence of China and India as major coal importers has placed enormous pressure on the international coal market, compounded by the Chinese government's decision to limit coal exports. India's coal imports have increased five-fold between 2003 and 2011. China's coal imports have increased 17-fold during the same period; and China's exports to Japan have dropped to less than 15 per cent of their 2003 level (MoF 2013). Competition from new Asian importers coupled with the emergence of a seller's market over the past decade, has resulted in Japan being faced with unprecedented coal prices in a sellers' market despite its dominant import position (IEA 2008: 100). Due to the increase in coal demand in Asia and the coal export infrastructure shortage in Australia coal prices have increased significantly over the past decade, raising concerns in Tokyo. Coal prices tripled in 2007 due to heavy rain in Queensland and increased by over 50 per cent in December 2010. In the absence of any new sources of large-scale supply, the Pacific coal trade's reliance on Australia was emphasized by the Queensland floods of 2010–2011, which saw coking coal spot prices soar to US$385 (Bowden 2012: 20). The average price per ton of Japan's thermal and coking coal imports peaked in 2008, coincidentally also the year in which Japan imported the highest volumes of coal. While the average import price dropped slightly since the 2008 peak, it still remains considerably higher than prior to 2008 (MoF 2013). Prices have partly remained high due to Queensland floods in early 2011 and again in 2013. In 2012, coal prices followed a slight downward trend for both thermal coal and coking coal largely due to the oversupply of coal in China and a decrease in US coal consumption for power generation due to the emergence of shale gas.

While Australia is perceived as a stable supplier, Japan's overreliance on Australian coal and the Australian dominance of the Asia–Pacific coal market, leaves Japan overly exposed to supply disruptions, such as during Queensland floods, but also during bilateral negotiations on long-term prices between Australian coal producers and Japanese utilities and trading houses. This is a particular issue for Japan regarding thermal coal imports, as Japan's demand is increasing and it relies on Australia for over 70 per cent of its imports. For example, some Japanese power producers, such as TEPCO and J-POWER will require more imported thermal coal in 2013 with the restart of the Haramachi Thermal Power Plant which has been shut since Fukushima and as new plants come online (Cooper 2013, IEEJ 2013a: 11). In 2012, Japan's thermal coal imports increased due to growth in demand for electricity generation, while coking coal imports have been flat due to lower steel production. Japan's six largest utilities expect to increase thermal coal consumption by 24 per cent in the fiscal year starting 1 April 2013 from the previous year in order to reduce demand for increasingly expensive LNG (Tsukimori and Tan 2013). If such an increase materializes, Japan's thermal coal imports are set to hit another record over the next year.

While in April 2013 Japanese utilities managed to negotiate a 14 per cent reduction in long-term prices with the Australian sellers over the previous year, largely due to lower Newcastle spot prices (Tsukimori and Maeda 2013) it is doubtful whether similar success will eventuate in subsequent annual negotiations. This is largely due to forecasted increase in coal demand in Japan and elsewhere in Asia. As further evidence of a sellers' market in operation, since 2007, Japan has consistently paid a higher price per unit of energy for Australian thermal and coking coal than for coal from other suppliers. Since 2009, on average, Japanese importers paid 10 per cent more for Australian thermal coal than for coal from other suppliers (MoF 2013). Japan's increasing demand for thermal coal and continued overreliance on Australia, where a few large coal producers dominate the market, will likely provide Australian producers with more bargaining power in annual bilateral negotiations with Japan, particularly given that there is an increased interest in Australian coal from China and India.

Greenhouse Gas Emissions Abatement and Continued Reliance on Coal

Coal is the largest source of power globally and its wide availability and relatively low cost imply it will stay so for the foreseeable future. However, coal electricity generation causes higher GHG emissions when compared to other fuels and its major drawback is that it creates a relatively large amount of greenhouse gases. Carbon dioxide emissions from coal-fired plants are about twice that from LNG-fired plants. The future of coal use largely depends on the pace of actions to mitigate climate change and on the development of new technology to minimize GHG emissions from its use (Rubin 2013: 37). The widespread deployment of more efficient coal-fired power plants is an essential first step for the longer term use of coal. This needs to be a high priority, especially where power plant fleets are being rapidly expanded. If the global average levels of efficiency of coal-fired power plants were 5 per cent higher, such an accelerated more towards more efficient technologies would lower global GHG emissions from the power section by 8 per cent (Birol 2013).

In addition, CCS technology is expected to play a major role in delivering economy-wide CO_2 savings over the coming decades. The deployment of CCS technology on a significant scale is a potential 'game-changer' for coal as it is the only technology available to drastically reduce GHG emissions from fossil fuels that allows the world to reap their benefits without the negative impacts associated with climate change. If widely deployed, CCS technology could potentially reconcile the continued widespread use of coal with the need to reduce GHG emissions. The CCS technology should deliver roughly 20 per cent of the mitigation effort required to stabilize global temperature increase at 2°C. To achieve CO_2 intensity factors that are consistent with halving CO_2 emissions by 2050, deployment of CCS is essential (IEA 2012). However, CCS has yet to be demonstrated on a large scale in an integrated fashion in the power and industrial sectors, so costs remain uncertain (Birol 2013: 9). The current picture is still one in which many legal, regulatory and economic issues need to be resolved. A particular challenge is to move forward with the dozen or so global demonstration projects that integrate these technologies.

More specifically, the process of capturing CO_2 is energy intense and results in higher energy consumption per output of energy produced – an energy penalty. As a result, capturing CO_2 at coal-fired power plants reduces the overall operating efficiencies of the plant in the range of 8 to 10 per cent compared to standard plants. It is vital to increase average energy efficiency of coal-fired power plants in order to minimize the energy penalty and to reduce capital and operating costs of CCS-fitted power plants. Given that CCS technologies are not being developed and deployed quickly, the importance of deploying advanced coal technologies to reduce emissions from coal-fired power plants is even greater in the medium term. High-efficiency-low-emissions coal technologies are a key element of the global effort to reduce CO_2 emissions, offering a low-cost opportunity for addressing the challenge of climate change. Climate change policy frameworks should incentivize a more efficient use of coal as one of the lowest cost opportunities for reducing greenhouse gas emissions (World Coal Association 2012).

While Japan has been committed to a reduction in its GHG emissions since the Kyoto Protocol was signed in 1997, a decrease in Japan's GHG is highly unlikely without a gradual decrease of the coal use or further improvements in coal-fired power plant efficiency and CCS technologies (Gasparatos and Gadda 2009: 4046). The issue is that Japan's reliance on coal seems to reflect energy security concerns that will not change in the foreseeable future and have been exacerbated by the Fukushima disaster. In the aftermath of the disaster, energy (and electricity) supply concerns override environmental concerns and this is unlikely to change in the near future. As a result, the prominence of coal in Japan's energy mix, particularly for electricity generation, is likely to either remain unchanged or increase. This will present a significant barrier to conformance with the Kyoto Protocol targets and with the even bolder targets of any follow-up strategy.

Prior to the Fukushima disaster, the MoE enforced a ban on the construction of new coal-fired thermal plants based on its environmental assessment, a measure it considered necessary to tackle climate change. However, as discussed above, in April 2013, the Japanese government encouraged the utilities to move to coal-fired plants, since coal is a relatively cheap energy source. TEPCO is likely to cut its oil purchases by more than one-third as it boosts its reliance on coal plants to reduce an energy bill that's ballooned since the Fukushima nuclear crisis started. From April 2013, TEPCO will generate or buy as much as 54 per cent more electricity from coal-fired plants compared with 2012, according to calculations based on company statements. That may enable it to reduce its purchases of crude and fuel oil by as much as 68,000 bpd (*Bloomberg* 2013). With recent government move to allow construction of new coal power plants, coal is expected to play an important role in the future. At the same time, Japan will have to deal with environmental issues, such as CO_2 emissions from its use. Therefore, efforts to make use of coal will continue to be integrated with the development of clean coal technologies.

While there are challenges ahead, Japan is at the forefront of technological advancement in making coal less emissions-intensive. Technological progress has been made in curbing CO_2 emissions at coal-fired plants in Japan, but more ways to make up for such drawbacks must be found. Since coal-fired power generation will likely continue to be a leading power generation method in Japan, Japanese companies are working towards reductions in emissions from coal-fired power generation. Japan Coal Energy Center (JCOAL) has been commissioned by METI and New Energy Development Organization (NEDO) to carry out various surveys into technology that uses coal, such as collecting coal-related data from all over the world. JCOAL, which has over 100 major public listed companies as members, promotes commercialization of clean coal technologies developed in Japan and transfers and disseminates these technologies to countries that produce and consume coal in the Asia–Pacific region. JCOAL also exploits and promotes coal projects to address global environmental issues (Tamaru 2010).

In order to contribute to reducing environmental impact from coal usage, JCOAL and the Japan Iron and Steel Federation developed ground-breaking coke-manufacturing technology, which can cut CO_2 by 20 per cent compared with conventional methods. Applying this technology not only reduces CO_2 emissions, but by blending to 50 per cent of it with steam coal, it also helps Japan in dealing with the soaring cost of coking coal. It is forecast that this technology will be commercially available in several years, making possible an 18 per cent reduction in the cost of coke production (JCOAL 2013).

Moreover, JCOAL has developed a high-efficiency hydrogen production process from coal. The fundamental concept behind the process is the integration of a water-carbon reaction, water-gas shift reaction and CO_2 absorbing reaction in a single reactor, which allows for increased carbon capture. JCOAL are also developing ashless coal (hyper coal). For a combined cycle power generation system to burn it directly with a gas turbine, JCOAL expect to achieve a net efficiency of 48 per cent. The amount of hyper coal by means of solvent extraction is about 60 per cent of the original coal and the remaining 40 per cent of the coal becomes residual coal with an ash content of about 15 per cent. This residual coal can be used in existing thermal power systems with pulverized coal combustion. The overall efficiency when using both the hyper coal and the residual coal is expected to be 45 per cent and the emissions of CO_2 could be reduced by 13 per cent compared with existing coal-fired thermal power generation (JCOAL 2013).

JCOAL are also working on a series of joint demonstration tests with Australia using oxy-fuel combustion at an existing coal-fired power station (Callide A power plant in Australia) to capture CO_2 and inject and fix it underground. This is the world's first demonstration test of the CO_2 CCS technique that sequesters CO_2 from a coal-fired power station. The demonstration test was selected as a Flagship Project of the Asia Pacific Partnership on Clean Development and Climate. When the CCS technique is applied, coal-fired power plants will use oxygen to burn coal instead of air which is usually used. Most of the flue gas discharged from this combustion system is CO_2, which can be easily separated and captured. For this flue gas, smaller SO_x and NO_x removal systems are needed or can even be eliminated. The technique allows low-cost CO_2 recovery in both existing and new power plants and CO_2 storage in coal beds or underground aquifers. If successful, this project will achieve clean coal-fired power generation free of CO_2 (JCOAL 2013).

Moreover, Hitachi has developed the world's leading supercritical pressure and ultra-supercritical pressure coal-fired power generation technologies, which are expected to contribute to the reduction of CO_2 emissions by achieving more efficient coal-fired power generation. The efficiency of coal-fired power generation is primarily dependent on when the steam generated from the boiler and the combustion of coal is at a more elevated temperature and pressure. However, the strength of the boiler decreases when used at high temperatures and pressure for long periods of time. In an effort to solve this problem, Hitachi has reviewed the design by focusing on strength and heat transmission and developed high-strength steel. Hitachi has also established ultra-supercritical pressure power generation technology, which is able to withstand high temperatures and high pressure. This approach has resulted in the reduction of CO_2 emissions by 7 per cent in ultra-supercritical pressure power generation, compared with current sub-critical pressure power generation. Moreover, Hitachi is applying a practical application of coal gasification power generation and the development of oxygen-burning coal-fired thermal power.

A handful of countries, including Japan, have made it a priority to improve the efficiency of their coal fleets. Coal-fired power generation in Japan is operated with a total efficiency rate of 42–43 per cent, the highest rate in the world (World Coal Association 2012, IEA 2012). This is due to the recent widespread utilization of supercritical pressure and ultra-supercritical pressure power generation, developed by companies such as Hitachi. In fact, with over 70 per cent of electricity generation capacity from coal in Japan from power plants with supercritical and ultra-supercritical capacity (IEA 2012), Japan already has the most efficient and least environmentally damaging coal-fired thermal power technology in the world (Tamaru 2010, Ando 2012). It is plausible that over the next decade efficiencies for ultra-supercritical coal plants without CCS could reach 50 per cent, so that the CO_2 emission rate for such plants would be 35 per cent less than for the global average coal plants in 2009 (Williams 2013: 54). Given the emphasis on further improving efficiency of its coal fleet, Japan is well positioned to be one of the first countries to reach the 50 per cent efficiency.

Unit 2 at J-POWER's Isogo Thermal Power Station is an exemplar for low emission coal-fired plants. The second unit at the plant entered commercial service in July 2009. The 600 MW ultra-supercritical Unit 2 joins an earlier, similar plant built in 2002. Together, these two new plants replaced 1960s-vintage coal-fired plants and doubled power generation from the small project site. In addition, the new unit improves the plant's gross thermal efficiency to about 45 per cent. Isogo ranks as the cleanest coal-fired power plant in the world in terms of emissions intensity, with levels comparable to those from a natural gas-fired combined-cycle plant. This technology is available for 'repowering' Japan's existing coal plants.

Conclusion

This chapter examined two major challenges related to Japan's coal use: the overreliance on Australia for imports and Japan's recent activity in overcoming the obstacles towards less emissions-intensive usage of coal. Historically, Japan has been highly dependent on the Asia–Pacific region and particularly Australia for its coal imports. Since China's transformation from a net coal-exporter to the world's largest coal importer over the past decade, Japan's reliance on Australia has increased. At the same time, major Australian coal producers have gained an upper hand in the international coal market, which leaves Japan at risk in bilateral negotiations. Moreover, high reliance on Australian thermal coal subjects Japan to the risk of supply disruptions, such as recurring Queensland floods. While Japan's utilities have been remarkably successful in reducing prices for thermal coal imports from Australia in 2013, the government's new policy to allow construction of coal-fired power plants and the accompanying bullish prediction for Japan's future thermal coal consumption may leave Japan vulnerable in forthcoming bilateral negotiations with its Australian suppliers. In addition, this chapter demonstrates that if coal is to continue to play an important role in Japan's future energy mix, Japan has to continue to aggressively support R&D into CCS and high-efficiency coal-fired thermal plants, such as Isogo. In fact, the chapter illustrates that Japan is already one of the world's leaders in these technologies. If the government remains committed to GHG emissions reduction in its future energy policy, successful development of these technologies will be crucial.

Chapter 6
Nuclear Energy

Introduction

Japan, the only nation to have been devastated by nuclear weapons, has been able – remarkably – to transform itself over the post-war period to a nation that has one of the largest civilian nuclear programs in the world (Lesbirel 2003). Japan's policy-makers originally chose nuclear power as a strategic necessity in order to enhance national energy security, buffer the economy from energy shocks and perhaps even serve as an important export product (Kim and Byrne 1996). Japan's historic commitment to a robust nuclear energy program is largely due to its lack of domestic fossil fuel resources, which has resulted in a heavy dependency on imported and costly oil, natural gas and coal. As a country with a poor resource base, Japan has historically viewed nuclear power as a major pillar in its longer-term energy strategy which is designed to reduce dependence on imported oil by developing alternative energy resources (Lesbirel 1990: 267). As a reliable semi-indigenous power resource, nuclear energy also afforded Japan greater flexibility and became important for political, economic and security considerations.

One of the origins of Japan's ambitious nuclear policy lies in the concerns of Japanese leaders who have interpreted history as a series of unreasonable assaults on an island nearly devoid of natural resources. They perceive Japan as exposed to inexplicable supply disruptions and argue that Japan would be too weak without recourse to an independent energy supply (Samuels 1994). Nuclear energy has been an integral part of Japan's energy supply system. The benefits of nuclear energy for Japan have been manifold. Nuclear energy adds to energy diversification (Lesbirel 2004), reduces dependency on oil, can be produced at a stable price and is a clean fuel in terms of emissions.

With the successful development of its own nuclear industry, Japan has regarded nuclear power since its beginnings as a 'semi-indigenous' energy source (Suttmeier 1981: 107). Fast breeder reactors offer the possibility of reducing fuel dependence. The government has justified the transition from oil-fired power to nuclear power by reasoning that 'uranium can be considered a domestic energy source in view of the fact that it can be utilized for some years after importation' (ANRE 2006). Driven by high dependency on imported fossil fuels and negative impact of the two oil crises, the government has been committed to nuclear power as a preferable energy source because it is domestically produced and thereby more secure. To promote energy diversification since the oil crises, Japan continually used Middle East oil dependence ratios as a key rationale for seeking to maintain public support for nuclear power at the national level and in seeking to persuade local communities to accept nuclear plants (Lesbirel 2004: 21).

Further rationale for historical promotion of nuclear power in Japan has been that nuclear power generation is most resistant to the disruption of fuel supply. This resistance is due to the lead-time for fuel procurement that can be characterized as a feature of nuclear power generation technology. It takes about two years to mine, concentrate, process and charge a nuclear power plant with uranium. Even if procurement contracts should be disturbed, uranium fuel procured under old contracts will continue to arrive at the nuclear power plant for the next two years. Moreover, once the fuel is charged into the reactor, normally it does not need replenishment for up to one year, making it possible to use an average of half a year under normal operation. This lends credence to the idea that uranium offers strong resistance to the disruption of supply as compared with oil, which has a short lead-time and needs constant replenishment. Consequently, when Canada chose temporarily to suspend uranium exports due to concerns over nuclear non-proliferation after India's nuclear test in 1974, Japan was not affected (Suzuki 2000b: 17). Moreover, uranium, the fuel source, is widely located in stable countries, such as Canada and Australia, thus promising a stable supply.

The biggest impact of an oil crisis on Japan's economy has not been caused by the physical shortage of the fuel itself, but by abrupt fluctuations in price. In contrast to oil prices, nuclear fuel costs have been relatively stable. During the oil crises in the 1970s, there were times when uranium prices soared. Nevertheless, uranium fuel costs account for less than 10 per cent of total nuclear power generation costs. Thus, if uranium fuel prices were to double, nuclear power prices would only rise by up to 10 per cent. As the nuclear power reactors operate past the period of paying off their capital costs, they produce electricity cheaper than any other energy source, because their operations and maintenance costs are the lowest (Nakata 2002: 364). The cost of a unit of electricity generated at a nuclear power plant was estimated as the lowest when compared with the cost of electricity produced using any other traditional or renewable source of energy. According to Matsuo et al. (2011), between 2006 and 2010, nuclear power generation cost less than thermal power generation when fossil fuel prices soared. They have demonstrated that while thermal power generation is directly affected by fluctuations of the fuel cost accounting for more than 70 per cent of the total cost, nuclear power generation is invulnerable to fossil fuel price fluctuations.

Moreover, Japan's government has promoted nuclear power as it perceives it is one of the few options that can help countries meet base load electricity demand with virtually no GHG emissions. Consequently, as a non-fossil fuel that does not generate GHG emissions, nuclear power has been considered a trump card in reducing Japan's GHG emissions. The 1998 interim report of a Demand and Supply Sub-committee Meeting of the Advisory Committee for Energy stated that for Japan to achieve its goals set forth in the December 1997 Kyoto Protocol, it will be necessary to increase nuclear power generation capacity from 4,500 kilowatt-hours (kWh) to nearly 7,000 kWh. This recommendation has led to an energy policy to 'build new nuclear power plants' (Suzuki 2000b: 19).

Consequently, in complying with the goals of both the Kyoto Protocol and the METI's long-term energy strategy (energy security, environmental protection and economic growth), nuclear power has played an important role prior to the Fukushima disaster. Prior to the Fukushima disaster, 25–30 per cent of Japan's power came from its nuclear energy program, which was slated to scale up to 50 per cent by 2030 (METI 2010a). It has been suggested that while Japan cannot solve its energy problems only with the utilization of nuclear power, it can never solve the problems without such a useful energy source as nuclear power (Ikuta 1980: 28).

This strong rationale in favour of nuclear energy in Japan has been challenged by the magnitude of the Fukushima disaster. Japan's nuclear history is full of monopolies and lavish subsidies, of cosy business–government relationships behind closed doors and of plans and targets that fall somewhere between 'bold and ambitious' and borderline fantasy. The conduct of the industry has been marked by error and malpractice, data falsification, the concealment of incidents and the denigration of risk (Tolliday 2012: 1). Most importantly, the nuclear village has been in control of the policy and regulatory processes throughout Japan's civilian nuclear history and it has been suggested that malpractice and lax regulatory oversight are one of the main culprits for the Fukushima nuclear accident. Consequently, the first challenge discussed in this chapter pertains to analysis of regulatory capture in Japan's nuclear industry and the power of the nuclear village. More specifically, will Japan's reformed nuclear regulatory structure be independent of industry capture?

Even prior to the Fukushima disaster, nuclear accidents, policy contradictions, NIMBY protests and regional funding issues have presented recurring obstacles to the government's plans to perfect a completely indigenous fuel source for the twenty-first century (Scalise 2004: 168). The public opposition to nuclear power has grown significantly following the Fukushima disaster. However, public opposition to nuclear power following past accidents has returned to general ambivalence towards nuclear power within years following these accidents. Will Japan's public opposition to nuclear power diminish over the next few years? Finally, the principal energy policy decision for Japan to make over the next several years is about the future of nuclear energy. Will Japan return to business-as-usual and restart most of the reactors over the coming years, or will it phase out nuclear power? Prior to engaging these pressing challenges for Japan, the first section outlines the evolution of nuclear policy in Japan. The second and third sections analyse nuclear sector organization in Japan and the evolution of the historical nuclear supply–demand balance.

Evolution of Policy

The availability of cheap hydroelectric power in the 1950s and cheap oil in the 1960s gave Japan little incentive to embark on major programmes for nuclear power generation. Without a nuclear weapons programme and an established lobby of nuclear scientists, Japan was not under the same technological and political pressures as the nuclear weapons states to develop civilian nuclear power (Surrey 1974: 213). Nuclear power development in Japan began with the 1953 historical speech *Atoms for Peace* by US President Eisenhower at the United Nations. Immediately thereafter, in 1954, the *Atomic Energy Law* was promulgated and enacted in 1955, with the aim of ensuring the peaceful use of nuclear technology in Japan. Democratic methods, independent management and transparency were (and are) the foundation of nuclear research activities, as well as the promotion of international cooperation. Several nuclear energy-related organizations were established in 1956 under this law to further promote development and utilization, including the Japan Atomic Energy Commission (JAEC), tasked with developing *Long-Term Plans for the Research, Development and Use of Nuclear Power*; the Science and Technology Agency (STA); the Japan Atomic Energy Research Institute (JAERI) and the Atomic Fuel Corporation (renamed the Power Reactor and Nuclear Fuel Development Corporation in 1967). Initially, Japanese electric power utilities purchased designs from US vendors and built them with the cooperation of Japanese companies, who received licenses to then build similar plants in Japan. Companies such as Hitachi, Toshiba and Mitsubishi developed the capacity to design and construct LWRs (Scalise 2004: 168).

While the institutional development mirrored that of the United States, soon afterward, however, Japan departed from America's primarily market-based approach to energy policy. Rather than allowing private energy utilities throughout the nation to handle the issues of siting and public acceptance on their own, the Japanese government developed an extensive array of policy instruments and social control techniques designed to bring public opinion in line with national energy goals. Authorities and regulators overcame opposition and concerns among the broader population and in specific demographic groups, such as coastal fishermen and students, through focused policy instruments intent on manipulating public support (Aldrich 2012a: 3).

The government provided a number of different types of support to TEPCO and other regional power monopolies in the early years of nuclear power. One form of help involved logistical and financial support in mapping out potential host communities throughout Japan. Government bureaucrats assisted the utilities both in the physical charting of potential locations – to ensure that they met certain technocratic criteria, such as having access to cooling water, proximity to existing power grid lines, support from relatively aseismic rock and so forth – and in mapping the social characteristics of nearby communities. Internal documents from the Japan Atomic Industrial Forum industry group showed that planners of the late 1960s and early 1970s were very cognizant of the dangers posed by

well-organized horizontal associations, especially fishermen's cooperatives. Analyses of the siting of nuclear power plants in Japan demonstrate that planners placed these projects in rural communities, which were less coordinated and more fragmented and, hence, less likely to successfully mount anti-nuclear campaigns. To overcome any remaining opposition in such localities, the government often offered jobs and assistance to fishermen to ensure that the nuclear power plant would not be seen as curtailing their livelihoods (Aldrich 2012a: 3).

The high and unstable price of oil – critical for Japan's petrochemical industries, as well as a host of other fields, including automobile production and oil refining – created pressure for Japanese planners to achieve a new goal: energy security. The government hoped that between hydroelectric dams and nuclear power plants, Japan would be able to wean itself off oil from the Middle East. This would require the consent of the citizens of Japan on a large scale. As a result of this new push, the system that allocated benefits to actual and potential nuclear power plant host communities became so complex that the central government created a new agency, the ANRE. Over the course of the next decade, more spin-offs were created, including the Japan Atomic Energy Relations Organization, the Japan Industrial Location Centre and the Centre for the Development of Power Supply Regions (Aldrich 2012a: 3).

Since its introduction in the 1960s, nuclear power has played an important role in Japanese energy policy, which has aimed to diversify energy sources in order to provide an insurance cover against unexpected oil supply interruptions and price increases. In response to the two oil crises in the 1970s, nuclear power has been elevated to a central place in energy policy. As outlined in the introduction, to Japan, nuclear power has a special attraction over other alternative energy sources. It has been regarded as a semi-indigenous form of energy. Fuel rods only need to be changed every couple of years and there is the prospect of completing the nuclear fuel cycle and developing fast breeder reactors, which will allow Japan to generate its own nuclear fuels (Lesbirel 1990: 268).

On 1972 costs, nuclear power was 16 per cent more expensive per kWh than oil-fired generation, but the dramatic increase in oil prices since October 1973 has reinforced the judgement that nuclear power would gradually become economic (Surrey 1974: 213). In the 1970s, the introduction of nuclear power generation began to accelerate. After the oil crisis in 1973, nuclear power became the major electricity source as an alternative to oil. In 1974, three electric power laws were promulgated (the *Law for the Neighbouring Area Preparation for Power Generation Facilities*, the *Electric Power Development Promotion Law* and the *Electric Power Development Promotion Special Accounting Law*). Under these laws, electric power location subsidies were to be given to the municipalities (prefectures, cities and towns) that agreed to accept nuclear power generation and other large-scale power generation plants. Nuclear power plants were to be given subsidies twice as high as coal-fired or oil-fired thermal power plants, providing a powerful financial incentive. The *Electric Power Development Promotion Tax* was incorporated into electricity bills to fund these subsidies and this ensured that

subsidies and research funds played a major role in promoting Japan's policies of developing energy alternatives to oil, especially nuclear power (Suzuki 2000b).

Re-evaluation of domestic energy policy following the first oil crisis resulted in diversification and in particular, a major nuclear construction program. A closed fuel cycle was adopted to gain maximum benefit from imported uranium (World Nuclear Association 2012). Nuclear power generation was perceived as largely insensitive to fuel price fluctuations. This factor and the perceived ease of long-term stockpiling of uranium, as well as the forecasted improvements in fuel-use efficiency have reinforced nuclear power's pivotal role in fuel substitution of electricity generation (Nemetz 1984/1985: 572). As a consequence, Japan saw its nuclear power development make steady progress in the 1970s and the 1980s. The number of nuclear reactors at existing sites has increased steadily through continued expansion since the late 1970s. Nuclear power has maintained an image as a low-cost, stable source of electric power in Japan, with the nine major electric power companies all owning nuclear power plants by the late 1980s.

The Three Mile Island nuclear plant accident in Pennsylvania in 1979 sent shock waves across the Pacific. There were temporary shutdowns in Japan to check equipment, but the drive for more nuclear plants continued. Developing nuclear energy was considered essential for any long-range hope to liberate the country from its dependence on foreign oil (Klein 1980: 44). However, since the Chernobyl accident in April 1986, the three electric power laws have proven less effective in gaining new sites for nuclear power, with the exception of Totsu, in Aomori prefecture. Opposition from rural residents, local governments and national anti-nuclear citizen movements has made it difficult to find sites for nuclear power plants since the Chernobyl accident. In fact, since the 1980s, the major impediment to the achievement of Japan's longer-term nuclear goals has been the siting and public acceptance issue (Lesbirel 1990: 267).

In 1988, the government placed the emphasis on nuclear power development in its annual White Paper. The 1988 White Paper by the JAEC, instead of focusing on the government's ambitious plans to reduce Japan's dependency on oil, stressed the safety of nuclear energy and the environmental benefits of not burning oil. It quietly dropped plans outlined in the 1986 White Paper to build 120 nuclear power plants by 2030. Nevertheless, in subsequent White Papers the government continued to maintain its commitment to expand nuclear power as Japan's main energy source (Dauvergne 1993: 581). In the early 1990s, MITI and STA were promoting plans to bring 40 new nuclear plants on line by 2010, thus increasing nuclear energy's share of total electricity output to 40 per cent.

A major research and fuel cycle establishment through to the late 1990s was the Power Reactor and Nuclear Fuel Development Corporation (PNC). Its activities ranged widely, from uranium exploration in Australia to disposal of high-level wastes. After two accidents and PNC's unsatisfactory response to them the government in 1998 reconstituted PNC as the Japan Nuclear Cycle Development Institute (JNC), whose brief was to focus on fast breeder reactor development, reprocessing high-burn up fuel, mixed-oxide (MOX) fuel fabrication

and high-level waste disposal. A merger of JNC and JAERI in 2005 created the Japan Atomic Energy Agency (JAEA) under the Ministry of Education, Culture, Sports, Science & Technology (MEXT), making it a major integrated nuclear R&D organization (World Nuclear Association 2012). Moreover, between 1975 and 2001, the government earmarked more than $2 billion annually for nuclear industry support. This commitment to nuclear power research allowed Japan to nurture the largest national per capita number of researchers, scientists, engineers and technicians in the nuclear power field (Sovacool and Valentine 2012: 116).

By the turn of the century, Lidsky and Miller (2002: 128) argued that even if energy security were not an issue, Japan's considerable investment in nuclear power argues against a sudden change in its long range plans to rely on nuclear power for a significant fraction of its electrical power needs. Massive 'locked-in' investments in nuclear power speak to the difficulty of rapid transformation in Japan's energy structure and the protracted nature of energy transitions. After 2000, increasing prices for fossil fuels and climate change policy targets have provided the most serious arguments in favour of nuclear energy. In March 2002, the Japanese government announced that it would increase its reliance on nuclear energy to achieve GHG emission reduction goals set by the Kyoto Protocol (Shadrina 2012: 73). In addition, in November 2002, the Japanese government announced that it would introduce a tax on coal for the first time, alongside those on oil, gas and LPG in METI's special energy account. At the same time METI would reduce its power-source development tax, including that applying to nuclear generation, by 15.7 per cent – amounting to ¥50 billion per year. While the taxes in the special energy account were originally designed to improve Japan's energy supply mix, the change was part of the first phase of addressing Kyoto goals by reducing GHG emissions. The second phase, planned for 2005–2007, was to involve a more comprehensive environmental tax system, including a carbon tax. These developments, despite some scandal in 2002 connected with records of equipment inspections at nuclear power plants, paved the way for an increased role for nuclear energy (World Nuclear Association 2012).

One of the specific numerical targets in the 2006 NNES was to increase the percentage of electricity generated by nuclear power to the level of 30–40 per cent or more from the level of 30 per cent (METI 2006a). Japan's 2010 BEP targeted the nuclear share of power production to surge from approximately 30 per cent to 50 per cent by 2030 (METI 2010a). The 2010 BEP contained much more ambitious and detailed targets for nuclear power than the 2006 NNES. The 2010 target represented an increase in the share of nuclear generated electricity of at least 10–20 per cent over that contained in the NNES. In addition and, in contrast to the NNES and its associated *Nuclear Energy National Plan* (METI 2006b), the BEP contained very specific figures for the number of new nuclear power plants to be built and the level of capacity utilization to be attained.

More specifically, the 2010 BEP called for the construction of at least 14 more nuclear reactors by 2030, with a combined capacity of 18 GW, assuming none of the 54 existing reactors were to be decommissioned in the meantime. Installed capacity of nuclear power was expected to reach 68 GW by 2030 to supply sufficient energy and reduce GHG emissions (METI 2010a). To promote nuclear power generation, the government aimed to extend the time between routine power plant inspections and to shorten shutdowns during inspections. Specifically, the plan was to increase the operating cycle from 13 months maximum between inspections to 18 months or longer by 2030 (IEEJ 2010: 6). It aimed to improve the power source location subsidy system, which it used to gain acceptance by local authorities and communities for nuclear power facilities. It would also take steps toward the establishment of the complete nuclear fuel cycle, including the development of pluthermal LWRs, which can use plutonium fuel and fast breeder reactors.

As demonstrated above in the survey of Japan's historical nuclear policy, since the beginning of Japan's nuclear power development effort, nuclear power has been regarded as 'domestic energy'. Given that since the 1973 oil crisis, a top priority of the Japanese government's policies has been development of alternative energy sources to oil, nuclear power has contributed significantly as an alternative to oil. However, following the Fukushima accident, in October 2011 the government published the White Paper on Energy Policy confirming that 'Japan's dependency on nuclear energy will be reduced as much as possible in the medium-range and long-range future' (METI 2011a). In July 2011, Enecan was set up by the cabinet office as part of the National Policy Unit to recommend on Japan's energy future to 2050. It was chaired by the Minister for National Policy and its focus was on future dependence on nuclear power. In early July 2012, through Enecan, the Japanese government presented three energy mix options in order to facilitate national debate: the zero scenario (nuclear power 0 per cent and renewable energy 35 per cent), the 15 scenario (nuclear power 15 per cent and renewable energy 30 per cent) and the 20–25 scenario (nuclear power 20–25 per cent and renewable energy 30–25 per cent). From mid-July to early August 2012, the government invited the public to send in comments, with about 89,000 comments received in total. According to a survey by the *Nikkei Shimbun* in August 2012, the zero nuclear power option gathered the most support with an overwhelming 76 per cent (IEEJ 2012f: 3).

Under the new laws that were passed in March 2012 (*Measures against Severe Accidents at Light Water Nuclear Power Reactor Facilities* and *Strengthening of Japan's Nuclear Security Measures*) regulatory standards and criteria for nuclear facilities were revised to become stricter and accident management is required by law. Previously, such measures have been provided by utilities on a voluntary basis. With new laws, comprehensive risk assessment of the safety design and operation of each reactor is also required. The new regulations are based upon cutting-edge findings and the latest knowledge on safety-related matters developed at existing nuclear facilities. In principle, operation of power reactors will be limited to 40 years; a certain period of operational extension could be approved on a onetime-only basis. Under the new regulations, nuclear operators will take responsibility for constantly improving the safety of their facilities. In this context, implementation of necessary measures for nuclear disaster prevention will be stipulated. There will be quality assurance requirements for a nuclear facility, not only in the operational stage, but at its design and construction stages (Shadrina 2012: 79). In June 2012 parliament amended the 1955 *Atomic Energy Basic Law* to stipulate that nuclear plant operators must prevent the release of radioactive materials at abnormal levels following severe accidents and that the NRA is to formulate regulations to achieve this. This law was initially promulgated before serious nuclear accident in Japan or globally and, consequently, the legislation did not include significant regulation on responses to accidents.

Enecan's *Innovative Energy and Environment Strategy* was released in September 2012, recommending a phase-out of nuclear power by 2040. In the short-term, reactors currently operable but shut down would be allowed to restart once they gain permission from the incoming NRA, but a 40-year operating limit would be imposed. Reprocessing of used fuel will continue. Enecan promised a 'green energy policy framework' by the end of the year, focused on burning imported LNG and coal, along with expanded use of intermittent renewables. This provoked a strong and wide reaction from industry, with a consensus that 20–25 per cent nuclear was necessary to avoid very severe economic effects, not to mention high domestic electricity prices. The *Keidanren* said the Enecan phase-out policy was irresponsible and the head of the LDP agreed (World Nuclear Association 2012).

Consequently, four days after indicating general approval of the Enecan plan, the Cabinet backed away from it, relegating it as 'a reference document' and the Prime Minister explained that flexibility was important in considering energy policy. The timeline was dropped. Reprocessing used nuclear fuel would continue and there is no impediment to continuing construction of two nuclear plants – Shimane and Ohma. It was suggested that a new basic energy plan will be decided after further deliberation and consultation, especially with municipalities hosting nuclear plants (World Nuclear Association 2012).

The newly elected LDP government under Shinzo Abe is unlikely to commit to a nuclear phase-out, particularly after its Upper House election victory in July 2013, which gives it control over both houses of parliament. The government will allow restarting nuclear reactors that have been deemed safe by the newly formed independent regulator, the NRA. Initially, the NRA, which was established in September 2012, announced that it will decide the outline of the new safety standards by January 2013 and, following necessary adjustments, will draw up the draft new standards, the revised nuclear disaster preparedness guidelines and the *Regional Plan for Disaster Prevention* by March 2013. The new safety standards were finalized in July 2013, and the NRA started reviewing the power plants for restarting based on these new standards. With the utilities applying for 14 reactors to be restarted, in November 2013 the NRA announced that it has no fixed schedule to complete safety checks at idled nuclear power plants, possibly delaying reactor restarts and the supply of cheaper energy the Abe government wants to drive economic growth. In addition, the new METI Minister Toshimitsu Motegi stated that once experts' opinions were collected, the new government would take 'a major political decision' on whether or not to allow construction of nine reactors that currently exist only at the planning stage. The LDP election victory also implies that Enecan is now defunct.

Sector Organization

Until 2012, the Nuclear and Industrial Safety Agency (NISA), situated within MITI/METI, was responsible for nuclear power regulation, licensing and safety and conducted regular safety inspections of nuclear power plants until it too was disbanded in 2012. The STA was responsible for safety of test and research reactors, nuclear fuel facilities and radioactive waste management, as well as R&D, but its functions were taken over by NISA in 2001. The JAEC, under the authority of the Prime Minister's Office, was tasked with setting nuclear policy, promoting research and development and implementing nuclear energy projects. The Nuclear Safety Commission (NSC), a more senior government body established in 1978 under the *Atomic Energy Basic Law*, was responsible for formulating policy in collaboration with the JAEC until it was dissolved in 2012. Both were part of the Cabinet Office. Gradually, MITI expanded its influence on matters of nuclear safety, leaving the NSC only to review MITI's actions. In addition, MITI gained control over a portion of the research and development of nuclear power, control of which previously had fallen under the purview of the STA. MITI (reorganized into METI in 2001) focused on the promotion of nuclear policy and worked with the nuclear power industry to implement the JAEC's nuclear policy.

METI has been criticized that as a renowned heavyweight, who crafted Japan's industrial might, it has been directly involved with nuclear regulation for the purpose of utilizing technological advancements in the nuclear sector to trigger the development of the Japanese economy's other segments (DeWit 2011, Matanle 2011, Moe 2012). Most of the critiques contend that it is hard to expect the agency whose mandate is to promote the industry to act as an impartial safety watchdog. It also has been pointed out (Funabashi and Kitazawa 2012) that significant harm was caused by placing NISA, an administrative body tasked to regulate nuclear power safety, under the METI umbrella. The on-going structural reform aims to separate nuclear regulation from nuclear promotion and centralize regulatory duties into one agency, the NRA (Shadrina 2012: 77).

As part of the structural reform, on 31 January 2012, the Cabinet adopted a bill envisaging several steps in the direction of safety regulation reorganization (METI 2012a). The document called for creation of a special nuclear safety agency, which would be established as an external organ of the MoE by separating the nuclear safety regulation section of NISA from METI and unifying relevant functions of other ministries. On 20 June 2012, the Diet enacted a law to establish the NRA (as a ministerial level agency unifying all safety regulatory activities hitherto executed by various ministries/agencies) from September 2012 (Shadrina 2012: 77). In September 2012, NISA and the NSC were disbanded and replaced by the NRA, which is modelled on the US Nuclear Regulatory Commission. The NRA also has the monitoring functions of the MEXT and has its own independent staff of 480 (mainly consisting of former NISA and MEXT officers) and a budget of ¥50 billion. Also coming under the remit of the MoE is a new five-member Nuclear Safety Investigation Commission (NSIC), which replaces the NSC and is tasked with reviewing the effectiveness of the NRA and responsible for the investigation of nuclear accidents. The NRA is expected to act independently; its top official, the NRA commissioner, has the authority to make an administrative decision on nuclear safety regulation, appoint and dismiss officials, make recommendations to relevant government organizations to ensure nuclear safety and draw on an independent account in the national budget. Regulatory and safety-related functions previously implemented by other ministries, such as regulation of research reactors, environmental radiation monitoring, radiation protection in emergencies and matters relating to nuclear security are now united in NRA. Crisis management is expected to be one of the most important roles of the NRA (Shadrina 2012: 78). The agency's first task is to make a pronouncement on the recommissioning of nuclear power reactors following the July 2013 release of regulations for plant system designs and severe accident management procedures.

All regional utilities in Japan currently operate nuclear plants with the exception of the Okinawa Electric Power Company. Nuclear vendors, such as Mitsubishi Heavy Industries, Toshiba, Hitachi, Japan Nuclear Fuel Limited and other companies, provide fuel in its fabricated form, ready to be loaded in the reactor, nuclear services and/or manage construction of new nuclear plants. Trading firms, such as Itochu, Marubeni, Sumitomo, Mitsui and Mitsubishi, coupled with electric utilities, are tasked with international uranium procurement. In Kazakhstan, Itochu signed an agreement to purchase 3000 tonnes of uranium (tU) from Kazatomprom over ten years in 2006 and, in connection with this Japanese finance, contributed to developing the West Mynkuduk deposit in Kazakhstan (giving Sumitomo a 25 per cent share and KEPCO a 10 per cent share). In 2007 Japanese interests led by Marubeni and TEPCO acquired 40 per cent of the Kharasan mine project in Kazakhstan and will take 2000 tU per year of its production. A further agreement on uranium supply and Japanese help in upgrading the Ulba fuel fabrication plant was signed in May 2008. In March 2009 three Japanese companies – KEPCO, Sumitomo and Nuclear Fuel Industries – signed an agreement with Kazatomprom on uranium processing for Kansai plants. In Uzbekistan, a Japan–Uzbek intergovernmental agreement in September 2006 was aimed at financing Uzbek uranium development and in October 2007 Itochu Corporation agreed to develop technology to mine and mill the black shales, particularly the Rudnoye deposit and to take about 300 tU per year from 2007. In February 2011 Itochu signed a ten-year large-scale uranium purchase agreement with NMMC (World Nuclear Association 2012).

In Australia, Mitsui joined Uranium One's Honeymoon mine project in 2008 as a 49 per cent joint venture partner. In early 2009, a 20 per cent share in Uranium One Inc was taken by three Japanese companies, giving overall 59 per cent Japanese equity in Honeymoon. In July 2008 Mitsubishi agreed to acquire 30 per cent of Western Australia's Kintyre project. In February 2009 Mega Uranium sold 35 per cent of the Lake Maitland project to the Itochu Corporation (10 per cent of Japanese share). The remaining 90 per cent of the Japanese interests in Lake Maitland project is owned by Japan Australia Uranium Resources Development Co. Ltd. (JAURD), which is acting on behalf of KEPCO (50 per cent), Kyushu Electric (25 per cent) and Shikoku Electric (15 per cent). In Namibia, in March 2010, Itochu acquired a 15 per cent stake in Kalahari Minerals. Kalahari owns 41 per cent of Extract Resources, which is developing the Husab project. In July 2010 Itochu bought a 10.3 per cent direct stake in Extract giving it 16.43 per cent in the project (World Nuclear Association 2012).

Supply–demand Balance

Japan has no indigenous uranium. Its 2011 requirements of 8195 tU were met from Australia (about one third), Canada, Kazakhstan and elsewhere. Prior to the Fukushima disaster, Japanese companies were taking equity in overseas uranium projects. As outlined in the previous section, Japan has secured mainly long-term contracts for uranium with a number of countries in order to diversify import sources (METI 2010b). The historical emphasis in the area of uranium supply was upon Japanese participation in overseas resource development, coupled with source diversification to ensure supply stability and economy (Nemetz et al. 1984/1985: 573). Japan's uranium suppliers, such as Australia and Canada, have largely been seen as more reliable over the long run than oil producers (Suttmeier 1981: 107).

Prior to the Fukushima disaster, electricity supply in Japan was highly dependent on nuclear power, which provided 25–30 per cent of electricity demand and 13–14 per cent of overall energy demand, from a total of 54 nuclear power plants. Historically, Japan's demand for nuclear power increased significantly since the 1970s oil crises, peaking in 1998 (Figure 6.1). Following the peak, Japan's nuclear power demand has dropped marginally until the Fukushima disaster. The most notable drop was experienced in 2003 due to safety checks on TEPCO's 17 reactors following revelations that the company failed to report a series of accidents.

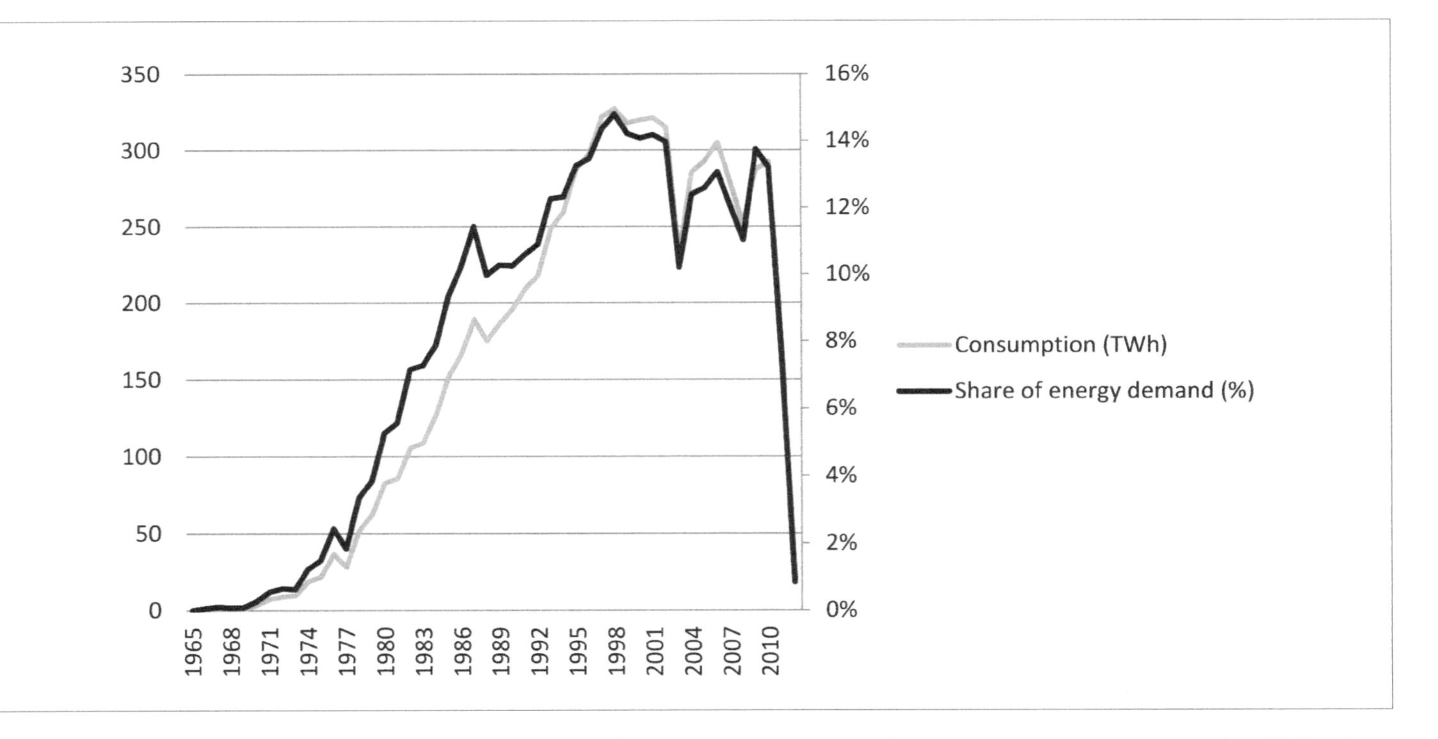

Figure 6.1 Japan's nuclear power demand (left axis; TWh) and share of overall energy demand (right axis) (1965–2012)
Source: BP 2013, IEEJ 2013b.

Following the Fukushima disaster, in mid-May 2011, only 17 out of Japan's 50 remaining nuclear power reactors (apart from written-off Fukushima Daiichi 1–4) were in operation. This represented 35 per cent of the total remaining nuclear generating capacity. Since then the number of operational reactors steadily dwindled to zero, with Japan *de facto* nuclear-free between May and July 2012. In mid-April 2012, after a series of high-level meetings, the Japanese government approved the restart of KEPCO's Ōi 3 and 4 reactors and urged the Fukui governor and the Ōi mayor to endorse this decision. The two reactors restarted in July in order to tackle looming electricity shortages in the Kansai region during summer. However, in September 2013, these reactors were shut down for scheduled maintenance. In 2012, nuclear power provided approximately 2 per cent of Japan's electricity supply and less than 1 per cent of overall energy supply (BP 2013; IEEJ 2013b).

Regulatory Capture and the Power of the 'Nuclear Village'

As briefly discussed in Chapter 1, in Japan, there exists a web of implicit threads woven into a fabric of nuclear energy politics. The network of a public/private 'policy community' involved with the promotion of nuclear power has often been referred to as the 'nuclear village' (Shadrina 2012: 73). Historically, policy and institutions for nuclear power in Japan have been insulated from everyday politics. They have been framed within a highly legalistic framework that has embodied large scope for bureaucratic discretion in policymaking and implementation. Everything from R&D to safety to commercial applications has been regulated through specific legal frameworks and institutional structures based on the *Atomic Energy Basic Law* and its revisions, with powers allocated to various government agencies that have codified, formal and legal powers to regulate safety, licensing, liability, compensation and so forth. The key advisory bodies have reported directly to the Prime Minister (not the Diet) and the key personnel in the agencies has been directly appointed by the Prime Minister, who has been obliged by law simply to 'fully respect' the recommendations of the bodies he had appointed (Donnelly 1993: 186–87). There has been consensus on aims since the 1950s and tight-knit personal contacts have been reinforced by *amakudari*, revolving doors between government and business and shared interests and ideas (Cohen et al. 1995, Dauvergne 1993, Lesbirel 1990).

Despite the existence of an overarching legal framework in the form of the *Atomic Energy Basic Law*, Japan's regulatory institutions have been subject to regulatory capture. According to Laffont and Tirole (1991: 1089), '"capture" or "interest group" theory emphasizes the role of interest groups in the formation of public policy'. In another interpretation, regulatory capture is the process through which regulated monopolies end up manipulating the state agencies that are supposed to control them (Bó 2006). In the case of Japan, institutional weaknesses including a lack of bureaucratic turnover and the failure to incorporate new ideas and ways of thinking, a lack of innovation and dynamism and a lack of transparency and accountability to both the fourth estate and the public, left the nuclear regulatory agencies open to capture.

Importantly, relationships between the bureaucracy governing nuclear power and electric utilities were – and still are – reinforced by the practices of *amakudari* and *amaagari* (Cohen et al. 1995, Dauvergne 1993, Lesbirel 1990). *Amakudari* ('descent from heaven') is a practice that sees retiring senior bureaucrats secure advantageous positions in the private or public sector (Colignon and Usui 2003, Blumenthal 1985, Johnson 1974, Mizoguchi and Van Quyen 2012). In contrast, *amaagari* ('ascent to heaven') sees industry members successfully gain employment in the regulatory agencies (Schaede 1995, Horiuchi and Shimizu 2001). *Amakudari* and *amaagari* are an omnipresent phenomenon in the electric power sector, with all major listed utilities having at least one former career bureaucrat sitting on the board of directors or serving in another role (Scalise 2012a). The practices are particularly prevalent in the nuclear power industry and it has not been uncommon for individuals involved in the nuclear sector to act at different times in the licensing, rulemaking and inspections process. For example, four former senior officials from nuclear regulatory agencies served as vice presidents of TEPCO between 1959 and 2010. Moreover, since 2000, electric utilities have supplied at least 100 employees to the NSC and other nuclear safety regulatory agencies. TEPCO, which has sent 32 workers to the bureaucracy, had reserved seats at several positions. Among other examples, a director general of the METI obtained a job with TEPCO after leaving his regulatory post. In addition, 68 former industry ministry officials with extensive nuclear industry oversight roles have transitioned to post-retirement positions as executive board members or advisers at 12 of the major power companies over the past five decades. As of May 2011, there were still 13 former ministry officials employed at TEPCO, with a further ten ensconced at other utilities (Shadrina 2012). In the most recent case, Toru Ishida became a senior adviser at TEPCO in January 2011, less than six months after retiring as the head of ANRE (Onishi and Belson 2011, Wang and Chen 2012). Private nuclear power industry companies may also have a direct role in shaping regulations. When the government convened a panel to revise nuclear regulatory standards in 2005, 11 of the 19 members on the panel were from the nuclear industry (Shadrina 2012).

In a further demonstration of the power and reach of the vested interests in determining policy and regulating the nuclear industry, *Associated Press* examined the business and institutional ties of 95 employees at NISA, JAEC and the NSC. Twenty-six employees in the sample were found to be affiliated either with the industry or with taxpayer-funded organizations that promote nuclear power. The media agency also identified 24 employees who previously had held positions at the three regulatory agencies, one-third of whom had connections to industry or pro-nuclear groups (Pritchard et al. 2011). However, perhaps no other person illustrates the movement and impact of *amakudari* and *amaagari* better than Tokio Kano. Kano joined TEPCO in 1957, became a leader in the utility's nuclear unit in 1989 and, in 1998 was elected to Japan's House of Councillors as one of the LDP's handpicked members of the *Keidanren* (Onishi and Fackler 2011). In the Diet, Kano participated in redrafting the policy that enshrined nuclear power as Japan's best hope for an energy secure future. After two six-year terms, he returned to TEPCO as an adviser in July 2010 (Scalise 2012a, Wang and Chen 2012).

With such incestuous relationships between nuclear regulators and nuclear utilities, it is unsurprising that inspections of nuclear power plants lacked rigorous regulatory oversight and preparedness. For example, despite Japan's regulatory documents listing the Fukushima Daiichi plant as one of the country's most trouble-prone reactors during the previous decade, NISA permitted its continued operation and, in February 2011, approved Unit 1 for a ten-year extension (Kaufmann 2011, Tabuchi et al. 2011). From autumn 2002 to mid-2003, TEPCO closed all 17 of its nuclear reactors as a consequence of falsified reports in which the company concealed scars on the shrouds or supporting devices of fuel rods inside the reactor. This situation suggested negligence in safety and security by TEPCO (Yokobori 2005). After TEPCO was found to have fabricated repair reports, the maximum fine that companies could receive for fraudulent reporting was raised to ¥100 million. However, TEPCO did not incur any sanctions as a result of its actions. Instead, the company sacked four top executives; ironically, three of these executives subsequently gained employment at companies with close ties to the utility company (Wang and Chen 2011). In the aftermath of the Fukushima disaster, the government largely left the response up to the plant's operator, TEPCO, which demonstrates a cosy relationship between government and the utilities. TEPCO has a track record of safety cover-ups, helped by soft regulation by a government organization tasked with promoting nuclear power.

In response to the string of nuclear accidents in Japan in the 1990s, experts called for a more adversarial regulatory culture and for the development of more appropriate laws and institutions. They also called for an effective nuclear safety and regulatory commission, which would be independent, transparent and encourage public participation (Kral 2000). Despite these calls, NISA remained responsible for nuclear power regulation, licensing and safety (World Nuclear Association 2012). NISA was not an independent regulator given its susceptibility to outside influence. However, NISA was not alone in exercising poor nuclear energy safety regulation: the JAEC and STA would regularly downplay safety considerations after consultations with METI. In the area of energy planning, the principal objectives and targets for energy policy in the METI's *Long-Term Energy Supply and Demand Outlook* would customarily converge with those in the JAEC's long-term plans for nuclear energy development. In matters of financial backing, the MoF would allot the funding for domestic research on and development of the same technologies that were readily available overseas. Flaws in administrative and regulatory routine would reproduce themselves in a new round of staff rotation among and between various 'districts' of the nuclear village (Shadrina 2012: 74).

The fact that the Japanese government did not restructure its nuclear regulatory framework since the accidents in the 1990s is an indication of the strength of the nuclear village. To a significant extent, it appears that regulatory capture of NISA and other regulators by Japan's nuclear industry turned the regulators into caretakers of industry rather than public safety providers. The concept of an embedded vested interest structure is central to understanding Japan's nuclear energy policy in the past. Safety, democracy and openness were the fundamental principles of Japan's nuclear energy policy when the country decided to diversify its energy sources in the 1950s. But these basic premises were undermined by the vested interests that controlled policy administration and implementation as the nuclear energy industry developed.

The failures in the nuclear regulatory system, including capture by industry and the government's attendant negligence in effectively exercising regulatory authority over the industry, were partly to blame for the Fukushima nuclear disaster (Wang and Chen 2012). As a consequence and, in response to public pressure, in September 2012, the NSC and NISA were abolished and replaced by the NRA. The former proved to be almost useless in advising the government during the hardest period of the Fukushima crisis, while the latter was too intractably involved with promotion of nuclear commerce and too ineffectual in its control of nuclear safety activity. Both not only had institutional and organizational deficiencies, but also were rather poorly staffed and incapable of providing expert opinion and guidance on the contingencies of Fukushima (Shadrina 2012: 78).

To ensure the new agency's independence from any governmental body in charge of nuclear power promotion, a 'no return rule' has been enforced in the agency's staffing policy. This rule means that top senior officials who join the new agency on loan from METI/NISA and MEXT will not be allowed to return to their original ministries. However, there are concerns about whether the NRA is not a mere nameplate change and window dressing. According to Kingston (2012d), the NRA is more a reorganization than a significant reform, as 460 of its 480 staff were transferred from NISA and the NSC. The NRA faces a number of challenges owing to its staff inheritance and changing the culture of how the regulator oversights the industry will be particularly important. The new NRA Chairman is Shunichi Tanaka, a former Vice Chairman of the JAEC. He also served as president of the Atomic Energy Society, an academic association that advocates nuclear energy. Because of his background, observers have expressed concerns about whether Tanaka will play a more robust monitoring role and whether regulatory capture will persist (Kingston 2012d). More broadly, it is important that the newly established body is free from the formerly exploited techniques and behaviour patterns. Without a doubt, the new system faces an enormous challenge to regain people's faith and demonstrate independence from the nuclear village in matters of nuclear safety and security (Shadrina 2012: 80). Moreover, it also remains to be seen whether the NRA will nurture a safety culture in the industry.

Public Opposition to Nuclear Power

Throughout most of the 1960s and 1970s, there was general consensus within the media, the general public and local governments in Japan on the necessity of nuclear power. Opposition was limited to a few people residing near nuclear plants. Although criticism arose briefly when pollution became a serious problem in the late 1960s and early 1970s, the oil shock in 1973 reinvigorated the drive for nuclear power as the need for energy self-sufficiency became a national concern (McKean 1981). In 1980 about 25 per cent of the Japanese public asserted that energy was its main security concern, in contrast with about 15 per cent in Germany and less than 10 per cent in France, the United States and the United Kingdom (Bobrow and Kudrle 1987: 546). During this period, it appeared as though local public opposition to nuclear development would not be a serious obstacle. Public opinion surveys during the 1970s indicated that nearly half of the people surveyed thought that nuclear power would be the principal source of energy in the future, while an equal number were worried about nuclear safety (Suttmeier 1981).

Although public acceptance of nuclear power was negatively affected by the Three Mile Island accident in the United States in 1979, three units of nuclear power stations were ordered in 1980 and another two units in 1981. This implies that, at the time, Japan was more accepting of nuclear power compared to the United States, the United Kingdom and West Germany, where almost no orders were made after the accident. Most likely this is because public acceptance of nuclear power has improved in response to the successive oil shocks of the 1970s. The Japanese public was well aware of Japan's overwhelming reliance on imported sources of fossil fuels and of the potential for events over which Japan would have little influence, which would threaten the country's energy lifeline with dramatic suddenness (Oshima et al. 1982: 96–97).

The April 1986 Chernobyl accident had a profound impact on nuclear sentiment. Although the accident did not directly lead to higher levels of radioactivity in Japan, the discovery of radioactive food imports from Europe generated tremendous concern among much of the Japanese public. Since then, the anti-nuclear movement has gained momentum, profoundly changing the context in which nuclear power has been developing in Japan (Dauvergne 1993: 578). In fact, growing opposition since the Chernobyl accident combined with technical and economic obstacles stalled Japan's drive for energy self-sufficiency through nuclear power. Prior to the Chernobyl accident, the Japanese media avoided discussing the dangers of nuclear power, especially compared with media attention to the issue in other countries. Just after Chernobyl, public opinion polls indicated for the first time that more Japanese people opposed nuclear power than supported it and, by late 1986, media coverage had become much more critical and grassroots opposition was growing rapidly. In April 1988, on the second anniversary of the Chernobyl accident, more than 20,000 people attended an anti-nuclear demonstration in Tokyo organized by about 150 local groups. According to a December 1990 government poll, 90 per cent of the public felt uneasy about nuclear power and 46 per cent thought that nuclear power in Japan was not safe. Although 65 per cent stated nuclear power was necessary, less than half supported expanding nuclear energy further (Deauvergne 1993: 580).

A series of minor mishaps in the late 1980s and early 1990s have contributed to the erosion of public confidence in Japan's nuclear industry. Japan's only nuclear ship, the Mutsu, was reactivated in April 1990, 16 years after its disastrous maiden voyage in which the reactor leaked and the engines crashed. The second try was no more successful. Initial attempts to launch the ship from a port near Aomori failed and the problems became so severe that, according to one report, 'even the nuclear officials in charge were frightened' (*The Economist* 1990: 34). The incident angered Aomori residents and added momentum to the growing anti-nuclear movement in that area. A series of mechanical mishaps at the TEPCO plant in Fukushima and the emergency shutdown of one of its reactors in late 1989 increased public concern and contributed to a strong antinuclear movement in the Fukushima area.

MITI and the STA responded to the growing concern over nuclear power by launching a public relations campaign in the late 1980s that has involved around 14,000 meetings between local groups, government and industry representatives and has necessitated a ten-fold increase in the government's public relations budget. TEPCO increased its public relations budget by 20 per cent. In October 1989 the government held the first *Nuclear Power Day* and 'posters showing a happy couple expressing their gratitude for nuclear power appeared in railway stations' (Cross 1988: 57). The government spent ¥4 billion in 1990 alone to encourage public acceptance of nuclear power (Dauvergne 1993: 581).

The growth in nuclear power in Japan throughout the 1970s and 1980s has occurred despite strong underlying concerns about risk. There is a strong nuclear allergy in Japan; it is the only country to have experienced nuclear devastation. The very term nuclear allergy (*kaku arerugii*) is used in Japanese civilian nuclear discourse not only by opponents of nuclear power, but also by proponents, reinforcing the importance of nuclear risk. Safety concerns have been heightened because Japan is earthquake-prone; several plants are located close to fault-lines and residents worried about the ability of containment vessels to withstand a large earthquake (Berger 1998). Public opinion polls showed high levels of sensitivity to the safety of nuclear power plants. For instance, a *Mainichi* newspaper survey (2 July 1989) showed that 80 per cent of those polled feared that nuclear accidents during the 1980s, similar to those at Three Mile Island and Chernobyl, could happen at any nuclear plant in Japan (Lesbirel 2003: 8–9).

In February 1991, Japan experienced its worst nuclear accident until that point in time, in the number two reactor at KEPCO's Mihama power plant. Although only a small amount of radiation leaked from its steam generator, the incident added momentum to Japan's anti-nuclear movement. Public trust in nuclear power was undermined for several reasons: Mihama plant operators were slow to alert the area of a problem, MITI and KEPCO officials at first denied that any radiation had leaked and, finally, subsequent investigation revealed that the cause of the accident was due to human error during plant construction, which had not been spotted during any of the numerous safety inspections (Buckley 2006). Even an official at ANRE, which has historically been responsible for the administration of Japan's nuclear plants, admitted that the accident will 'undoubtedly intensify concerns about nuclear power among the public' and will likely increase the problem of finding sites. Haruo Suzuki, director of the office of nuclear safety policy research at STA, stated that from a political point of view the accident was 'very severe'. He lamented that after the extensive public relations campaign to reassure the public in the wake of the Chernobyl accident, Mihama had 'blown a gaping hole in all those efforts' (cited in Swinbanks 1991: 644).

On 8 December 1995, the experimental sodium-cooled fast breeder reactor known as the Monju experienced a huge sodium leak. The resulting fire was hot enough to melt various steel structures in the chamber. The Japanese agency in charge of the Monju, however, decided to suppress details of the accident and to doctor a publicly released videotape of the leak and its aftermath. Local residents successfully fought attempts to restart the experimental reactor until the summer of 2005, when the Supreme Court ruled in favour of restarting (Aldrich 2012a: 5, Lesbirel 2003: 12).

The breadth of policy instruments for manipulating public opinion, while effective, has not guaranteed success at siting. By the late 1990s, siting planners encountered serious bottlenecks in the system of constructing new nuclear power plants. The time between the proposal of the plant and its activation stretched from less than a decade in the early 1970s to more than three decades by the late 1990s (Lesbirel 1998). Tatsuya Terasawa, former head of the electric power development division of ANRE, claimed in late 1991 that growing opposition in the 1980s was causing serious delays in Japan's nuclear program. Terasawa provided data showing that nuclear plants that were completed in the 1980s had taken an average of 17 years and four months from the time of their proposal to the start of operations. However, the 14 plants scheduled to begin operating in the 1990s had taken an average of 25 years and seven months from the time of their proposal – presuming they stay on schedule (Ako 1991).

Research has shown that of the roughly 95 attempts to site nuclear power plants in Japan over the post-war period, 54 were actually completed, with comparatively lower completion rates since the 1990s (Aldrich 2012). Citizen opposition to nuclear power because of potential health effects, the lack of a long-term storage facility for nuclear waste and potential proliferation concerns grew steadily. The Citizens' Nuclear Information Centre and the anti-nuclear newspaper *Hangenpatsu Shinbun* publicized ongoing fights against siting attempts and provided advice to would-be opposition groups. Across the industrialized democracies, residents began to demand more from their governments, moving beyond basic materialist concerns to focus on the environment, sustainability and health. In addition, a series of large- and small-scale accidents and cover-ups in the industry, including three fatalities at a nuclear facility in Tokaimura, chipped away at public support for the industry in the mid-1990s (Aldrich 2012: 5).

There has been a significant loss of confidence since the mid-1980s in the effectiveness of once highly regarded government institutions and public concern about nuclear risk increased quite dramatically during the 1980s and 1990s as a result of several nuclear accidents (Lesbirel 2003). The accidents at Three Mile Island and Chernobyl worried many Japanese residents, but authorities reassured them that these would not be possible in Japan, given its strong engineering credentials, in-depth safety controls and highly educated and motivated staff. Consequently, unlike most industrialized countries, opposition from outside forces has had surprisingly little impact on the overall direction of Japanese nuclear energy policy. The government also enlarged the range of projects to which the funds for local communities that accepted to host nuclear reactors could be applied, lengthened the period for which they would be available and increased the pool of funding provided to local communities. Overall, despite ongoing opposition, the government and regional energy monopolies saw few reasons to worry about the future (Aldrich 2012: 5).

MITI spent large and growing amounts on 'public acceptance' campaigns for new reactor sites after the Three Mile Island (1979) and Chernobyl (1986) disasters increased public resistance to reactor locations (Lesbirel 1998, Aldrich 2008, Dusinberre and Aldrich 2011). Early patronizing attitudes and highhandedness were quickly seen to be counterproductive and MITI developed a much more subtle and coordinated strategy of inducements, including 'economic development assistance' to communities that agreed to host new plants (sums continually ratcheted up as communities learned to drive cannier bargains). Accidents and public resistance did not make MITI/METI rethink its nuclear expansion targets; rather, they simply pressed ahead with aggressive expansion targets using the carrots of massive infrastructural projects and immensely sophisticated pro-nuclear public relations (Dusinberre and Aldrich 2011: 692–93). METI avoided confronting public opinion across Japan as a whole and instead concentrated on locating reactors in regions where resistance was least likely and could be most easily bought off, for example, in regions where reactors could be presented as 'depopulation countermeasures' (Aldrich 2008: 128–35). It avoided compulsory purchase or direct central control as far as possible. Instead it placed resources at the disposal of the electric power companies and left them to handle local negotiations backed up by its central guidance and support. The 'assistance' was funded partly from central government funds but also from a levy on the utilities. The utilities accepted this because they were then permitted to recoup this cost from their energy pricing – an important factor in persistent high consumer electric power prices. Despite differences among utilities, MITI/METI and their supporters in the LDP on other matters, they collaborated harmoniously in the face of outside challenges (Aldrich 2008, Dauvergne 1993, Donnelly 1993).

Opinion polls taken throughout the 1990s reflect uncertainty in the public over what stance to take on Japan's nuclear energy policy. A series of polls taken from 1990 to 1994 by the NHK Broadcasting Cultural Research Centre demonstrated that while support for building more nuclear power plants hovered between 7 and 11 per cent and support for abolishing nuclear power between 9 and 13 per cent, support for 'taking a cautious attitude' towards nuclear power ranged consistently between 50 and 60 per cent. A follow-up poll conducted by the NHK Broadcasting Cultural Research Centre showed results were essentially unchanged as those found in the earlier polls despite coming after nuclear power-related accidents at Monju in 1995 and Tokaimura in 1997. The 1998 poll found that only 8 per cent of respondents believed that it is necessary to build more nuclear power plants in Japan; 10 per cent supported abolition of nuclear power; while 55 per cent said it was necessary to take a cautious attitude. In a February 1999 poll conducted by the Prime Minister's Office, a majority of Japanese (69.9 per cent) supported the existence of nuclear power, but an almost equal majority (68.3 per cent) were worried about the effects of using nuclear power generation (Kotler and Hillman 2000: 21–22).

Among the series of reactor accidents in Japan during the late 1980s and 1990s (documented in Beder 2003, World Nuclear Association 2012), the most serious was the 30 September 1999 accident in Tokaimura (Lesbirel 2003: 12). The accident happened when three workers at the nuclear fuel cycle company JCO in Tokaimura were preparing fuel for one of Japan's experimental fast breeder reactors set off a criticality (an increase in nuclear reactions in radioactive material) that exposed them to tremendously high levels of radiation. Two of the three died from extreme radiation exposure and local residents in the nearby town were told to remain indoors to avoid contamination (Aldrich 2012a: 5–6). In fact, prior to Fukushima, the Tokaimura accident was the worst nuclear accident in Japan and it fundamentally shattered the trust of the Japanese people in the industry's management capabilities for nuclear power generation and affected the future of Japan's nuclear power industry (Suzuki 2000b: 21).

The 1999 Tokaimura accident has contributed greatly to negative public confidence in government and corporate nuclear oversight. The share of Japanese people feeling 'very uneasy' about nuclear power grew from 21 per cent before the 1999 Tokaimura accident to 52 per cent afterwards. In an October 1999 Japan Public Opinion Company survey, only 11 per cent supported government plans to increase the share of nuclear power, 51 per cent favoured maintenance of current plans, while another 33 per cent wanted to see a reduction in, or end to, nuclear power. In a survey released in March 2000 by the Japan Productivity Centre for Socio-Economic Development, 64 per cent of energy experts surveyed expressed strong concerns about the risk to energy security posed by limitations to securing sites for nuclear power plants; and 49 per cent about risks posed by large accidents at nuclear power facilities (JPC-SED 2000, Kotler and Hillman 2000: 19). This survey data shows that both the public and the experts did not accept the government's argument that nuclear power is safe well before the Fukushima disaster. In fact, the belief in a high degree of safety and trust in the Japanese nuclear power industry may have evaporated with the Tokaimura accident.

The minor nuclear accidents throughout the 1990s were not the only events that began to break apart public support and faith in the industry. Revelations that TEPCO had covered up numerous accidents, leaks and cracks since the 1980s also came to light. As discussed above, engineers came forward in the early 2000s to reveal that at least 30 serious incidents had been hidden by company management. In response, several upper management executives lost their jobs and the central government ordered the shutdown of TEPCO's 17 nuclear reactors in 2002. These events further undermined the industry's credibility (Aldrich 2012a: 6).

A series of minor nuclear accidents in Japan and, particularly the 1999 Tokaimura nuclear accident, slowed down the rate of subsequent nuclear development. The construction of new reactors has slowed greatly in the last decade because of safety concerns and local opposition. Actual and potential public opposition to new nuclear power plants, which could result in long delays in the licensing process, has made the utilities reluctant to invest heavily in them, given the high costs of construction. The 2006 NNES effectively called for an increase in the number of nuclear facilities, yet in 2010, all of the reactors planned in 2006 were three to five years further behind schedule. Thus, the JAEC concluded in 2009, 'No considerable growth is expected for the present regarding activities to construct new or additional plants in Japan' (JAEC 2009: 21). In 2001, the Japanese government planned to increase the number of nuclear power plants from 52 to between 62 and 65 by 2010 (Lesbirel 2003). Yet, by 2010, only two new power plants have been in operation.

Given that over a decade had passed since the Tokaimura accident, prior to the Fukushima accident, public perception in Japan towards nuclear power was relatively positive and had been improving: public surveys showed that the respondents who agreed with the addition of new nuclear power in the country reached 59.6 per cent in 2009, compared to 55.1 per cent in 2007 (Table 6.1). In a 2005 poll conducted by the International Atomic Energy Agency (IAEA), 82 per cent of Japanese respondents supported maintaining or expanding upon existing nuclear power capacity (Ramana 2011). Also, those who responded that they were unconcerned about nuclear power reached 41.8 per cent in 2009, versus 24.8 per cent in 2007 (Hayashi and Hughes 2013). Despite this, the Japanese government's aggressive nuclear policies have not been supported by all stakeholders; for example, since the 1990s, the string of nuclear accidents and scandals associated with electricity suppliers has significantly reduced local governments' trust in them as demonstrated by their opposition to the construction of new nuclear power plants (Sugawara et al. 2009).

Table 6.1 Change in public opinion on nuclear power (%)

Date of survey	Dec 2005 (1)	Oct 2009 (1)	Jun 2011 (2)	Oct 2011 (2)
Expand	55.1	59.6	3.0	2.1
Maintain status quo	20.2	18.8	24.4	23.2
Reduce and decommission	17.0	16.2	66.1	66.6
Don't know	7.7	5.4	6.5	8.2

Source: Miyamoto et al. 2012: 32; (1) published by the Public Relations Department of the Cabinet Office; (2) NHK Broadcasting Culture Research Institute.

However, since the Fukushima disaster, questions on nuclear power and power supply have become the subject of enormous public interest, with significant public perception shift. Initially, public opinion polls conducted in mid-April following the Fukushima disaster did not show a significant decline in support for nuclear energy, but over the next two months there was a significant shift against nuclear power in various polls. The late-May revelation by TEPCO that it had misled the public about the scale of the crisis by not acknowledging that there had been three meltdowns until then undermined public faith in the utility and nuclear energy (Kingston 2012a: 196–97). Consequently, within a few months after the Fukushima disaster, the public's opposition to nuclear power has become even more pronounced than following the Tokaimura accident. In the *Asahi Shimbun* poll of June 2011, 74 per cent of Japanese respondents favoured a gradual phase-out of nuclear energy and only 14 per cent were against such a gradual reduction. The poll also showed 64 per cent of respondents believed 'natural energy', such as wind and solar power, would replace nuclear power in the future (*The Australian* 2011). Public opinion polls done by the Roper Centre for Public Opinion Research in early August 2011 of some 1,000 residents across Japan reported that nearly 60 per cent of the respondents had either little or no confidence in the safety of Japan's nuclear power plants (Aldrich 2012a: 7). The transformation of public opinion toward nuclear power was an inevitable consequence of the crisis.

When Enecan proposed three future options for nuclear power in Japan – phase-out, 15 per cent and 20–25 per cent by 2030 – 76 per cent of the public, according to a survey by the *Nikkei Shimbun* in August 2012, supported the zero nuclear power option. However, other opinion surveys by the media show that the largest group of people is in support of maintaining nuclear power. More specifically, the *Asahi Shimbun* states that the zero scenario has the strongest support at 43 per cent while the *Mainichi Shimbun*, *Yomiuri Shimbun*, NHK and *Nikkei Shimbun* show that the largest group support the continuation of nuclear power at 64 per cent, 55 per cent, 54 per cent and 50 per cent, respectively (IEEJ 2012f: 3).

Following the Fukushima disaster, the prevailing sentiment among the Japanese has been that a move away from nuclear power towards other sources of energy is desirable. However, as noted above, public opposition to nuclear energy is not a new phenomenon. In the immediate aftermath of the Three Mile Island and Chernobyl disasters public opinion polls have registered voter opposition to nuclear power. This opposition increased following the series of domestic nuclear accidents in the 1990s, with the public signalling its opposition to Japan's ambitious nuclear expansion policy (Fesharaki and Hosoe 2011, Sovacool and Valentine 2012). Although opinion polls have shown movement towards the anti-nuclear spectrum following Fukushima, there have been few large-scale anti-nuclear rallies and host communities have not been overly hostile to the recommissioning of nuclear power plants (Aldrich 2012b: 136). Historical data on Japanese public opinion towards nuclear energy indicates that opinion has been quick to return from opposition to a state of general ambivalence following past incidents. With minimum focus on nuclear issues in the Japanese media through 2012 and into 2013, arguably, the government's aim has been to silence the nuclear power debate, while also hoping that the malleable public will return to its normal state of ambivalence during the course of the next few years (Interview with Paul Scalise, 16 January 2013, interview with Scott Valentine, 17 January 2013).

This is not a new tactic employed by the Japanese government and nuclear industry. Historically, government communication has been fairly successful in shaping public perception regarding Japan's energy policy challenges over the years and has achieved a state in which public perception and government policy are remarkably similar (Valentine et al. 2011: 1874). The government has continually forged ahead with ambitious plans to build more nuclear reactors, despite strong opposition from the public, media and local government. Historically, rather than altering plans, the state and industry have responded to growing criticism by launching expensive public acceptance campaigns (Sovacool and Valentine 2012: 120).

In addition, in the past and again following the Fukushima disaster, the Japanese government has utilized exclusive reporters' clubs (*kisha kurabu*) in order to ensure that media coverage reflects government policy. The continued existence of the discriminatory system of *kisha kurabu*, which restrict access to information to their own members, is a key element that caused the ranking of Japan's media freedom to drop from 22nd to 53rd since the Fukushima disaster (Reporters without Borders 2013). In fact, a respondent referred to the Japanese media as 'lap dogs of the politicians' (Interview with Benjamin McLellan, 11 January 2013). Japan has been negatively affected by a lack of transparency and the denial of access to information on subjects both directly and indirectly related to Fukushima. Following the Fukushima accident, there has been much evidence in Japan of censorship of nuclear industry coverage. The authorities also imposed a ban on independent coverage of any topic related directly or indirectly to the accident at the Fukushima Daiichi nuclear power plant. Moreover, several freelance journalists who complained that public debate was being stifled were subjected to censorship, police intimidation and judicial harassment (Reporters without Borders 2013).

More broadly, although investigation reports were issued by the government, the Diet and a private-sector committee as well as by TEPCO, a genuine account of the nuclear crisis has yet to be completed. Thus far, no one has been charged with crimes related to the nuclear power plant disaster. However, better access to TEPCO's records, as well as to what was known by affiliated companies and the government, is the first step to determine possible negligence, collusion or criminal responsibility. For that process to be fair and thorough, independent verification, one of the most important duties of a free press, is essential. The public deserves to know what the records of TEPCO and of the government might reveal. TEPCO has consistently barred access to documents and to people. When freelance and independent reporters were finally allowed into the plant, TEPCO demanded final say over their video and images. That does not constitute press freedom.

An investigative reporter was sued by one of TEPCO's subsidiaries to silence his reporting. Freelance journalists and magazines were sued after publishing articles on the alleged collusion among politicians, nuclear plant construction companies and TEPCO. Taking reporters and publications to court shows the intent to cover up the truth. In addition, in early December 2013, the Japanese parliament passed a state secrets protection law that may curtail future public access to information on a wide range of issues, including Fukushima (Sieg and Takenaka 2013), which has been condemned by critics of the Abe administration (Kojina 2013). Public officials and private citizens who leak "special state secrets" face prison terms of up to ten years, while journalists who seek to obtain classified information could be imprisoned for five years (McCurry 2013). These developments demonstrate the government's continued push to silence anti-nuclear dissent in the face of rising public opposition to the technology. In the past, Japan could be relatively proud of its reputation for press freedom compared with that of most countries. However, the government's and nuclear industry's tactics in which they aim to silence the public debate and prevent free reporting the nuclear disaster and more generally on Japan's future energy policy, question the very nature of Japan's democracy.

The Future of Nuclear Power

Over the past 50 years, Japan has developed one of the most advanced commercial nuclear power programs in the world. This is largely due to the government's broad repertoire of policy instruments that have helped further its nuclear power goals. These top-down directives have resulted in the construction of 54 plants and at least the appearance of widespread support for nuclear power. By the 1990s, however, this carefully cultivated public support was beginning to break apart. In 1999, Low et al. (1999: 81) argued that 'A major accident at a Japanese reactor would see the end of Japan's nuclear power program'. A major accident at Fukushima has raised and reinforced environmental concerns and health fears, as well as scepticism about information from government and corporate sources (Aldrich 2012a). It has also resulted in debate regarding the future of nuclear power in Japan.

The rational expectation was that the profound traumatic experience Japan endured on and following the 11 March 2011 natural disaster would divert the nation from the path of nuclear energy development. However, this anticipated outcome does not conform to reality (Shadrina 2012: 70). Some have suggested that the clout of the nuclear industry has been reduced by the Fukushima disaster (DeWit 2011, Kingston 2011) and that Japan needs to strive to phase out nuclear power (Valentine 2011). However, other analysts support the restart of the reactors as soon as possible, arguing that politicians are catering to popular sentiment and that this will set the country on the wrong course (Kasai 2012). For the economy striving to maintain its international competitiveness, production cost is a considerable issue.

Although the number of proponents of the nuclear energy cost-efficiency argument has declined considerably after the price of the Fukushima accident became known, there is still much support for nuclear energy based upon this very economic reasoning (Shadrina 2012: 72). With increasing costs of fossil fuel imports putting pressure on Japan's economy, the IEEJ (2013a), claim that it is important to steadily restart those nuclear power stations that are shown to be safe by the NRA. They have calculated that restarting 26 nuclear power stations – such as those that have submitted the results of stress tests – in 2014 would lower the electricity costs by ¥1.8 trillion and the electricity generation cost by approximately ¥2/kWh. Many changes outside Japan confirm the importance of nuclear power for Japan. One example is the rising tension in the Middle East. Iranian oil exports have dropped from the normal level of 2.5 million bpd to 1 million bpd due to the economic sanctions and the risk of travelling through the Strait of Hormuz has increased (IEEJ 2012i: 3).

When consideration of the policy implications of the Fukushima event began, there was a feeling among knowledgeable observers that no changes in Japan's future nuclear energy policy would occur quickly. The topic was considered too painful and sensitive to the general public, too dangerous and slippery for the career of any ranking government official or politician to receive serious immediate attention. It was clear that before any changes to nuclear energy policy could be made, reconciliation with the reality of nuclear energy safety and restoration of public confidence in the nuclear power industry were essential (Shadrina 2012: 69). While there appears to be consensus among energy policymakers in Japan that a greater level of oversight is required to ensure that Japan's nuclear power operators are held to higher standards of accountability, a technocratic ideology that is sympathetic to nuclear power development is still prominent (Sovacool and Valentine 2012).

According to Kingston (2012d), the deck is stacked in favour of the pro-nuclear advocates of the nuclear village and it is unlikely that public opposition will trump the networks of power defending nuclear energy. The marginalization of public opinion is evident in a number of significant policy developments since the Fukushima disaster. On 14 September 2012 the Noda Cabinet appeared to endorse a gradual phase-out of nuclear power by the late 2030s, but within days quickly disavowed this plan under heavy pressure from business lobby groups. Having muddied the waters on energy policy, the Cabinet then shifted responsibility for any future reactor restarts to the NRA and reactor hosting communities. The head of the NRA initially demurred, stating that his organization merely assesses operational safety and does not have responsibility for reactor restarts. He later recanted, but pointed out that utilities also share responsibility for reactor restarts because they have to get local communities to agree. Dispersing and blurring the lines of responsibility and authority is a strategy for undermining opposition.

On 18 September 2012, the Cabinet declared that the government would strictly adhere to the law on decommissioning reactors over 40 years old. However, later that day the Noda Cabinet said that decision is up to the NRA. The 40-year rule has a major loophole that allows renewal of the operating license at the discretion of regulators, the same ones tarnished by three major investigations that found that cosy and collusive relations between nuclear watchdog authorities and the utilities compromised safety in Japan's nuclear plants and was a major factor leading to the accident at Fukushima (Kingston 2012c). Given regulatory capture in Japan, meaning that nuclear regulators have long regulated in favour of the regulated, there are good reasons to doubt that stricter guidelines will be resolutely implemented. Extending the operating licenses for aging, old technology reactors may be dangerous, but profitable for the utilities (Ramseyer 2012).

The Cabinet caved on its pledge to phase out nuclear energy just one day after the nation's three largest business groups *Keidanren*, *Keizai Doyukai* (Japan Association of Corporate Executives) and the Japan Chamber of Commerce and Industry issued a joint statement complaining that the government had ignored their objections to a nuclear phase-out. *Keidanren* Chairman Yonekura Hiromasa inveighed, 'We object to the abolition of nuclear power from the standpoint of protecting jobs and people's livelihoods. It is highly regrettable that our argument was comprehensively dismissed'. Publically admonished by the nuclear village elders, the Cabinet promptly flip-flopped. This was a major victory for the nuclear village as a 'no decision' opened opportunities to lobby politicians and shape public opinion. Policy drift and biding time play to the advantages and interests of the nuclear village. The nuclear village was also hopeful that the LDP will regain power and resume support for nuclear energy as a centrepiece of national energy strategy (Kingston 2012d).

The US role in influencing Japan's government's nuclear policy cannot be discounted. Historically, US–Japan relations played a significant role in encouraging Japan to invest in nuclear power (Onitsuka 2012). A 'zero-nuclear' Japan is a serious concern for the United States as its key ally both from economic and security standpoints (Kitazume 2012). If Japan phases out nuclear power but continues to extract plutonium from fuel, it will raise proliferation fears, which Washington is strongly concerned about. The US also does not want Japan to stop using nuclear power due to the likely negative commercial effect this may have on America's export plans. Media reports suggested that Washington pressured Tokyo into fudging the Cabinet's official endorsement of the no-nuclear goal to provide some flexibility in future policy (Ito 2012). Consequently, this evidence suggests that due to its own commercial and security interests, the US indirectly supports the continued dominance of the nuclear village in Japan.

Another development suggesting that the nuclear village has prevailed is the government's decision to allow the completion of three new nuclear reactor projects that had been suspended following Fukushima. Now former METI Minister Edano Yukio argued disingenuously that the government has no authority to cancel licenses previously issued. Approving construction of three new reactors in the face of overwhelming public opposition to nuclear energy is a sign that pulling the plug on nuclear energy has been abandoned in favour of business as usual. In addition to the Oma plant where construction is about 37.6 per cent complete, one in Shimane is 93.6 per cent complete and may come online as soon as 2014. The other project in Higashidori is still in the initial stages and only 9.7 per cent complete; analysts doubt whether this project will proceed because the remaining financial hurdles are too high (Kingston 2012d, Ito 2012). Deftly moving to defuse public criticism over the new reactor construction restarts, Edano stated that the government is drafting a new law that would enhance its powers to veto the construction of new power plants. Currently, the NRA has legal authority over approving new reactor construction proposed by the utilities so the new laws might be a strategy for METI to claw back some power; previously NISA held authority for approving nuclear construction projects and was an agency within METI. METI wants the final say on licensing of new projects, but given that it controls subsidies paid to communities agreeing to host nuclear plants, it already has considerable leverage and theoretically could withhold such subsides if its wishes are ignored. Yet, given that former METI/NISA employees constitute a vast majority of the new NRA staff, the chances of major differences over nuclear energy issues are minimal (Kingston 2012d).

Although many expected that nuclear power would feature prominently in the pre-election campaign, it did not become a major campaign issue in December 2012 Lower House election. A shift in voter perceptions regarding the future role of nuclear power did not materially impact the political arena (Scalise 2012a: 148). While all parties other than the LDP pledged – albeit to varying degrees – to do away with nuclear power generation, the election outcome suggests that a large portion of unaffiliated voters who support an end to nuclear power voted for the LDP. In contrast, *Nihon Mirai no To* (Tomorrow Party of Japan), whose sole agenda was to abolish nuclear power, lost many of its pre-election seats (Sawa 2013). Given the unpopularity of the Noda government and overwhelming public support for the zero option, it is revealing that the DPJ has not played the anti-nuclear card to attract voters. While the DPJ nominally takes an anti-nuclear stance, its pre-election actions have been quite supportive of nuclear energy and not consistent with its pledge to phase out nuclear energy. This signifies that political leaders are more willing to risk public ire than defy the nuclear village (Kingston 2012d). The newly elected government under Shinzo Abe's LDP – historically an integral part of the nuclear village – will not commit to a nuclear phase-out. Shinzo Abe stated that nuclear phase-out is unrealistic and irresponsible (*ABC News* 2012).

Since the Fukushima disaster, the public pressure has altered decades of business-as-usual nuclear politics in Japan. Without a doubt, the Fukushima accident will inevitably hinder the future development of nuclear power in Japan (Sawa 2012: 126) and dampen nuclear power expansion ambitions for the short-term. It is quite certain that at least four of the six nuclear reactors at Fukushima Daiichi will be closed permanently and some the remaining reactors are unlikely to resume operation in the near future in light of public concern (Zhang et al. 2012b). While the pre-Fukushima strategy for national energy involved siting up to 15 more nuclear power plants over the next few decades, with the goal of increasing nuclear power's share of electricity production to 50 per cent, plans have clearly taken a new direction. With the establishment of NRA, the government has taken an important step in promising to separate nuclear regulators from nuclear promoters. While some have suggested public opinion will continue to be the determining factor influencing the nuclear policy (Bhattacharya et al. 2012: 34), history of past mishaps in Japan suggests that the current levels of strong public animosity may gradually fade. The key is that the nuclear village retains veto power over national energy policy and citizens may not get to decide the outcome (Kingston 2012d). The revitalization of Japan's nuclear power industry is likely once public sensitivities over the Fukushima crisis have died down (Sovacool and Valentine 2012: 131).

The evidence presented in this section indicates that Japan's nuclear power programme unlikely to be shut down. Why has Fukushima not been a game changing event? The institutions of Japan's nuclear village (principally the utilities, bureaucracy and Diet) enjoy considerable advantages in terms of energy policy-making. They have enormous investments at stake and matching financial resources to sway recalcitrant lawmakers and the public. The nuclear village has openly lobbied the government and actively promoted its case in the media while also working the corridors of power and backrooms where energy policy is decided. Here the nuclear village enjoys tremendous advantages that explain why it has prevailed over public opinion concerning national energy policy. Its relatively successful damage control is an object lesson in power politics. To some extent the lessons of Fukushima are not being ignored as the utilities are belatedly enacting safety measures that should already have been in place, but a possibility of a nuclear-free Japan by the 2030s remains uncertain.

Conclusion

Prior to the Fukushima disaster, nuclear power played an important role in Japan's energy mix, accounting for 25–30 per cent of electricity supply. Nuclear power was considered relatively safe, cheap and a domestic source of energy. However, following the Fukushima disaster, many in Japan have called for a nuclear phase-out, with public opinion strongly opposed to continued use of nuclear energy. The hitherto strong rationale in favour of nuclear energy in Japan has been challenged by the magnitude of the Fukushima disaster. In particular, regulatory capture has been cited as the main culprit for lax safety procedures that have indirectly lead to the nuclear catastrophe at Fukushima. This chapter demonstrates that the iron triangle of government bureaucrats, politicians and industry remains committed to the continued reliance on nuclear energy. In this context, it is argued that the jury is still out on whether Japan's reformed nuclear regulatory structure will be independent and free of industry capture. Moreover, given that public opposition to nuclear power following past accidents has returned to general ambivalence towards nuclear power within years following these accidents, it is possible that Japan's public opposition to nuclear power may diminish over the next few years.

Finally, while key members of the vested interest structure (or the nuclear village) remain in positions of power, some of the nuclear power reactors are likely to return to operation in 2014.

Chapter 7
Renewable Energy

Introduction

Historically, renewable energy has been on the sidelines of Japan's energy policy and demand. With hydropower largely exploited since the 1960s, 'new renewables', such as solar, wind and geothermal, have played a minor role in Japan's energy mix. With the exception of solar power since the 1970s, the renewables have also been on the margins of Japan's energy policy thinking, due to their relative cost and other structural issues, but also due to opposition from the electric utilities. In 2011, renewable energy provided around 8 per cent of Japan's total energy consumption and electricity generation, with hydropower accounting for three-quarters of the total. Since the 2010 BEP, Japan has been encouraging greater use of renewable energy for power generation. Most importantly, since Fukushima, in July 2012, the Japanese legislature approved a new FIT, compelling electric utilities to purchase electricity generated by renewable fuel sources at fixed prices. The costs are to be shared by government subsidies and the end users. The FIT has been adopted in order to increase the share of renewable energy in Japan's electricity supply.

A number of major challenges stand on the way to a greater penetration of renewable energy in Japan and they are the focus of this chapter. The chapter begins with a brief overview of the evolution of renewable energy policy in Japan, sector organization and supply–demand structure. The major challenges discussed in the chapter include institutional bias towards solar power, the potential of the new FIT and grassroots level support for renewable energy post-Fukushima to drive a major penetration of renewable energy in the future and various structural issues pertaining to various renewable energy sources that may inhibit their growth in Japan's energy mix.

Evolution of Policy

Japan developed hydropower rapidly in the first decade after the war and by 1965 some 40 per cent of Japan's electricity production came from hydroelectric plants (Nemetz et al. 1984/1985: 555). To increase the domestic supply of energy, the Japanese government launched an ambitious program in renewable energy sources. The Sunshine Project, initiated in 1974, focussed on developing solar, geothermal, coal conversion and hydrogen technology. Its original target was to provide 1.6 per cent of Japan's total energy supply by 1990 (Nye 1981: 215).

The Sunshine Project has been at the centre of Japan's efforts to develop clean, alternative energy sources to replace oil as the basic resource for Japan's energy needs. It has been described as an extensive technological development program projected on a very long-term basis and it has covered all areas of new energy-related technology with the exception of nuclear energy (Morse 1981b: 11–12).

In May 1980, the government passed the *Petroleum-Substitute Energy Promotion and Development Law*, which was designed to set goals for the development of new energy sources, investigate policies for commercial and technical applications (Morse 1981b: 11). In 1980, MITI created a subsidiary agency charged with developing alternative energies not related to nuclear power. The New Energy Development Organization (NEDO) was created to support projects focused on reducing Japan's oil dependency through technology development. NEDO, which began operation in October 1980, has concentrated its efforts on three core technologies – coal-to-liquids, geothermal and solar – while also providing domestic support to the promotion and dissemination of these technologies. In 1993, NEDO started a new, comprehensive research and development program on solar energy under the banner of the New Sunshine Project as a method of efficiently solving difficulties related to new energies, energy conservation and the global environment in order to address all these problems in a coordinated manner (Honda 1998: 115). The solar sector received support from NEDO from 1994 until 2002 and has reduced prices of photovoltaics by 60 per cent, making them affordable for many households (Stewart 2009: 181).

Consequently, Japan played a major global role in the development of solar energy from the late 1990s until several years ago. What has been lacking, however, is any substantial policy support for raising the share of renewables in Japan's energy mix, despite a promising government subsidy initiative for domestic PV systems that started in 1994 – a key factor that helped kick–start the solar industry. The subsidies died off from 2006 to 2008, leaving Japan trailing behind Spain and Germany in total grid-connected PV installed capacity (Bhattacharya et al. 2012: 9). More broadly, Japan's solar cell industry, which used to be the largest in the world, was surpassed in 2008 by Germany and China.

With pledges to cut carbon emissions by 25 per cent by 2020 from 1990 levels and, by roughly 80 per cent by 2050, the Japanese energy strategy pre-Fukushima largely focused on nuclear power as a nominally cheap, quasi-indigenous and low carbon power source. In contrast, renewables played a rather minor role before Fukushima. Historically, government funding for renewable energy has targeted specific technologies. For example, in response to the oil crises in the 1970s, solar thermal energy received an enormous boost in funding as the government sought to develop domestic energy sources. In the same period, funding for geothermal energy research also escalated as the government sought to expediently exploit Japan's abundance of geothermal sites. Some geothermal projects were installed in the 1970s and 1990s, but growth has slowed to nil since 1999 (Sugino and Akeno 2010). Throughout the 1980s and 1990s, the most consistent funding initiative was in solar photovoltaic (PV) research.

Annual government grants for solar PV research have exceeded US$50 million since 1981 (Valentine 2011: 6847).

By international standards, government-sponsored renewable energy R&D programs in Japan have been well funded (IEA 2008). However, the targeted funding approach applied over the past 30 years has produced winners and losers. Solar PV research has been by far the biggest winner. Although the cost of solar PV electricity has remained commercially unviable (about three times as expensive as wind energy), consistent funding has produced world-leading solar energy manufacturers. On the other hand, government funding in support of wind energy technology has been negligible. Aside from wind turbine prototype testing in 1992–1993 (Inoue and Miyazaki 2008), wind energy research has been largely ceded to private R&D initiatives (Valentine 2011: 6847).

Because solar and wind power are relatively high-cost and intermittent, support mechanisms such as feed-in-tariffs, green certificates, premiums and production tax credits are needed to induce residential and industrial support. The political support for and the inducement of policies towards increasing renewable energy production began long before the Fukushima disaster. With renewable energy drawing a great deal of attention in the series of discussions to determine post-Kyoto greenhouse gas emissions reduction targets, it was not completely neglected in Japan prior to the Fukushima disaster. In fact, there were two policy approaches for promoting renewable energy diffusion. The *Renewable Portfolio Standard (RPS) Law* (2003) was the first major policy in effect. In 1999, a collaborative effort between Diet members and the environmental NGO ISEP, led to an unsuccessful Diet initiative to introduce a FIT. The law was fought by METI and the utilities and never passed. Instead, METI proposed the RPS. The 2001–2002 political fight over the RPS was protracted and politicized, but ended with the utilities and the METI mainstream winning out against a coalition of METI's new energy bureaucrats and a number of Diet members (DeWit and Tani 2008, Maruyama et al. 2007). Under the RPS Law, electric power suppliers are obliged to supply a percentage of their customers' electricity demands through generation that utilizes 'new energy' sources. 'New energy' in this context means power from solar photovoltaic, wind, biomass, geothermal power that will not reduce hydrothermal resources, or small hydro power plants (less than 1 MW). The renewable target implied by this law was 0.39 per cent of total electricity sales by all distributors in 2003, 1.35 per cent in 2010 and 1.6 per cent in 2014. Until 2009, electricity distributors have over-achieved in meeting their obligations every year. Suppliers have been able to provide electricity with new-energy-fuelled generation capacity that they own, have been able to buy electricity from IPPs, or buy the equivalent value of 'CO_2-free' power by purchasing 'renewable energy credits' from other electricity suppliers (Takase and Suzuki 2011: 6736).

Given that until 2009, the utilities have had no problem in fulfilling the required electricity output, the main government instrument on renewables did not place any pressure or provided any incentives for the utilities. Some have suggested that the RPS was only introduced to pre-empt outside efforts at introducing a

FIT (DeWit and Tani 2008, Maruyama et al. 2007). The tug-of-war over the RPS caused profound consternation within METI as this was perceived as an outside rival for power (the Diet) seeking to wrest energy policy-making away from it. It led METI to become even more territorial about energy policy. Following the passing of the RPS it became politically very hard to put renewables back on the agenda (Iida 2010, Maruyama et al. 2007).

The eponymously named 'Fukuda Vision' was announced by the former Prime Minister following a speech entitled *Action Plan for Achieving a Low-carbon Society* on 29 July 2008 and in response to growing concern in Japan about climate change. Following this speech, the government implemented a series of policy measures to counter GHG emissions. Starting in January 2009, METI provided subsidies and tax credits for the installation and renovation of solar panels on residential homes (Scalise 2012a: 145). In late 2009, when it became apparent that the renewable energy goals for 2010 would not be met (the goal was 1.35 per cent of electricity production), the government established a FIT for surplus electricity generated by PV installed at residences and other entities, with a target of increasing the amount of installed PV capacity more than 20-fold (from 2.1 GW in 2008 to 28 GW) by 2020 (Duffield and Woodall 2011: 3747). A FIT guaranteed the power produced a fixed electricity purchase tariff for a specified period (often 10–20 years), typically in combination with preferential grid access for the electricity produced. The FIT forced electric utilities to buy surplus electricity from rooftop photovoltaic at twice the price than the 2009 market price offered under voluntary net-metering by the utilities. However, the scheme had been restricted to residential PV systems and rewarded only net electricity production of the households (IEA 2010a). The scheme was only applicable to PV installation below 500 kilowatt (kW) and to excess electricity generated (net metering).

A limited FIT was passed as it was never on METI's agenda. The utilities strongly opposed it and no major interest groups within METI perceived of any need for a change to the status quo (Iida 2010). With the LDP in power, METI usually got its way, largely because of its connections inside the party apparatus (Moe 2012). However, by 2008–2009 it became apparent that four decades of near uninterrupted LDP rule was ending. The DPJ came to power in September 2009, explicitly vowing to break up the iron triangle (*The Japan Times* 2009). This was causing great concern within METI. Thus, attempts were made to pre-empt the DPJ. One DPJ election promise was a general FIT for all renewables. However, in 2008, an LDP climate change policy group recommended the implementation of a FIT. The idea received general backing from the MoE as well as the other political parties. This took METI by surprise and it with extreme haste launched its own FIT proposal, as it did not want to lose control over the process. According to Iida and DeWit (2009), it was a 'half-baked scheme cooked up by METI's internal politics and client-list of vested interests', and a proposal constructed by people who were strongly against the implementation of any FIT. Whereas the DPJ plan was more comprehensive, the METI scheme gave almost exclusive preference to solar and applied to surplus power only, not gross power (Moe 2012). As a compromise, METI's scheme was eventually implemented in 2009.

In contrast to the 2006 NNES, which contained no specific targets for renewable energy, Japan's 2010 BEP contained the most ambitious and detailed targets for renewable sources of energy. Due to growing concerns about climate change, the government had subsequently (by 2009) established a goal of increasing the share of renewable energy in the primary energy supply from 6 per cent (2005) to 9 per cent in 2020 and 11.6 per cent in 2030. But the 2010 BEP raised these targets even higher, to 13 per cent. As per the 2010 BEP, to increase the use of renewable energy, the government would expand the 2009 FIT, which until then applied only to small-scale electricity generation by PV cells, to include wind, geothermal, biomass and small to medium-scale hydroelectric plants. The government would increase its support for the introduction of new renewable technologies, through such means as tax reductions, subsidies and support for research and development. It would take steps to deregulate the domestic energy market and prepare the power grid for intermittent sources of supply. Other measures that were considered by METI included introducing sustainability standards for biofuels and expanding the introduction of renewable thermal energy (METI 2010a).

Taken together, no stringent regulatory framework encouraging significant investment in renewable energy had been implemented before the Fukushima disaster (Moe 2012). Maruyama et al. (2007: 2763) even argued that Japan's renewable energy policy was impeding renewable energy use rather than contributing to its spread. According to Huenteler et al. (2012), the main reason is to be found in the regulatory structure. Most responsibilities related to the energy sector are concentrated in the METI, including the ANRE responsible for renewables. As established in Chapter 1, METI is by tradition 'in league' with powerful industry organizations, such as the *Denjiren* and the *Keidanren*. Through this 'cosy relationship' between monopolistic utilities, industry and a pro-nuclear bureaucracy, vested interests have been able to hollow out all attempts towards stringent policies for renewable energy (Aldrich 2011: 62, Moe 2012). Given the prioritization of enhancing economic security in energy policy planning, one perhaps should not be too surprised by the government's apathy toward renewable energy. Prior to the Fukushima disaster, all forms of renewable energy were considerably more expensive than nuclear and coal-fired power in Japan (Valentine 2011: 6848).

It has been proposed that in the aftermath of the Fukushima disaster Japan should follow the example of countries that successfully promoted the use of renewable energy, such as Germany (DeWit and Kaneko 2011, Iida 2011, Huenteler et al. 2012). Exploiting the potential of renewable energy should be accorded top priority if Japan wishes to minimize the externalities that reliance on fossil fuels and nuclear power cause (Valentine 2011: 6852). Former Prime Minister Naoto Kan had outlined in June 2011 a plan to increase the contribution of renewables to power supply to 20 per cent by 2020, a share that even the most ambitious plans did not envisage before 2030. To that end, he proposed to revise the existing FIT and to extend it to other forms of renewable electricity and offered his resignation conditional on, *inter alia*, the Diet passing it (Muramatsu et al. 2011). Amid the rising concerns about nuclear energy, a coalition of public figures, companies, politicians and non-profit organizations formed in support of the new bill (DeWit 2011). Eventually, on 26 August 2011 the extended FIT was passed by the Diet (the *Act on Purchase of Renewable Energy Sourced Electricity by Electric Utilities*, METI 2011b). The new FIT has been in effect since July 2012, encompassing PV, wind power, small hydroelectric, geothermal and biomass (Ayoub and Yuji 2012, IEEJ 2011c: 5). With the passing of new FIT, utilities are required to buy electricity from renewable energy providers at a rate of ¥42/kWh for solar energy, ¥23/kWh for wind power; ¥27/kWh for geothermal power; and ¥30–35/kWh for small scale hydro power (METI 2012c: 5).

Sector Organization

METI promotes renewable energy in Japan, assisted by other government agencies such as the National Institute of Advanced Industrial Science and Technology, the Institute of Applied Energy and the reformed and renamed New Energy and Industrial Technology Development Organization (NEDO). While METI and specifically ANRE is tasked with framing the direction of Japan's energy policy, which includes renewable energy, NEDO is Japan's largest public management organization promoting research and development as well as the deployment of industrial, energy and environmental technologies. NEDO defines its role as 'to foster research and development that individual enterprises are not capable of implementing alone'. NEDO publishes an annual plan of R&D funding programmes for which firms submit proposals. It also considers unsolicited proposals from firms and/or academics as a way of developing 'technological seeds', enhancing the national intellectual infrastructure and supporting medium to high-risk technologies that are 'difficult for private firms to address by themselves' (Harborne and Hendry 2012). There are no restrictions on foreign investment for participation in Japan's renewable energy sector. In fact, foreign investment is welcomed by the Government to help increase the country's renewable electricity base. As will be discussed below, government policy and support historically helped 'legitimize' solar PV as a major renewable energy technology for introduction in Japan.

SHARP, Sanyo-Panasonic, Kyocera and Mitsubishi Electric form the core of the Japanese solar panel industry. They also organized through the Japan Photovoltaic Energy Association (JPEA). Other important private actors include Toshiba, Mitsui, Honda, Solar Frontier and Fukuyama. Japan formerly had one of the largest programs in the world to promote rooftop solar panels. However, the domestic solar industry has since fallen somewhat behind the international market leaders. Their share in global PV cell manufacturing has dropped significantly since early in the past decade, when they were global leaders. The termination of Japan's solar incentive program in 2005 caused a decline in the local market.

Mitsubishi Heavy Industries and Marubeni are Japan's major wind turbine manufacturers and market participants, although they account for approximately three per cent of global wind turbine production. Japanese trading houses and industrial corporations have scaled up their international wind M&A activity since the Fukushima disaster. In September 2011, Marubeni announced the acquisition of a 49.9 per cent stake in the 172 MW operational Gunfleet Sands offshore wind farm in the UK for £200m. In similar fashion, Mitsubishi announced its intention to acquire stakes in the HelWin 2 and DolWin 2 German North Sea offshore wind transmission assets in March 2012 to add to its announced US$318m purchases of stakes in the BorWin1 and BorWin2 German offshore wind transmission assets a month earlier. In 2012, Mitsubishi has also acquired a 34 per cent stake in the 396 MW Mareña Renovable wind project situated in the state of Oaxaca, Mexico and a 50 per cent stake of the Walney 1 UK undersea offshore wind transmission assets. Japanese interest in overseas wind energy projects seems to be geared around learning how certain products and processes work with an aim of leveraging that knowledge in their future operations both in Japan and internationally (KPMG 2012).

More broadly, both solar panel and wind turbine manufacturers have highlighted the constraining influence of electricity utilities for delayed deployment of new energy technologies (Harborne and Hendry 2012). However, following the Fukushima disaster, the conservative and traditionally anti-renewable utilities have embraced renewable energy to some extent. Since the Fukushima disaster, KEPCO has stepped forward with new plans for safer alternative energy sources, including a new 10,000-kW solar facility in Osaka Prefecture. TEP has stated its intent to dramatically increase the capacity of its wind farms by 2020 (Aldrich 2012a: 7).

Supply–demand Structure

Japan has a hydropower capacity of approximately 30–35 GW. Much of this has been exploited via large hydropower facilities. According to some estimates there is about 12 GW of potential hydropower capacity at over 2,700 locations that may be technologically and economically feasible (Bhattacharya et al. 2012: 20, Takase and Suzuki 2011). Hydroelectric power accounts for approximately 70 per cent of Japan's overall renewable electricity production and hydroelectric production has remained relatively constant over the past four decades (Figure 7.1). Electricity production from other renewable sources has grown significantly since the 1980s, but has remained stagnant between 2005 and 2010. However, since the Fukushima disaster and the passing of new FIT, electricity production from wind and particularly solar PV has increased.

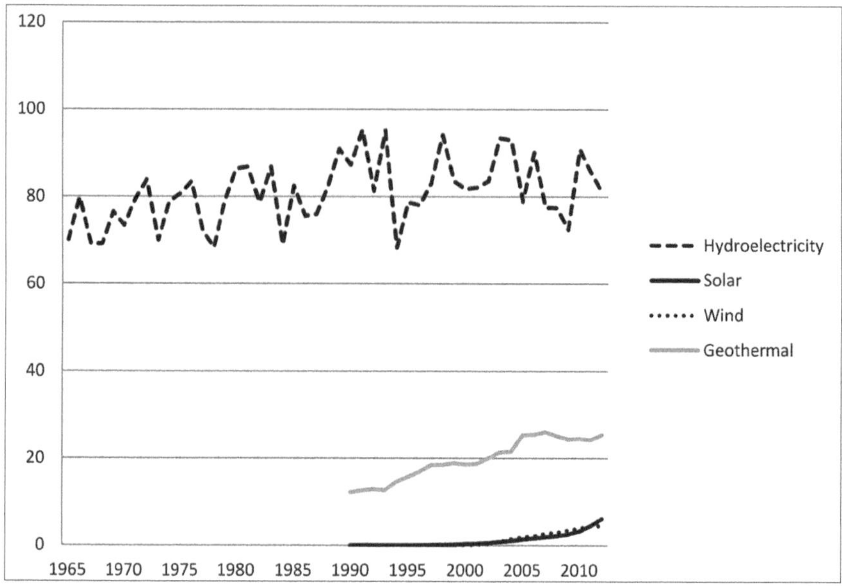

Figure 7.1 Japan's renewable electricity production (1965–2012; TWh)
Note: data for solar, wind and geothermal production is available only since 1990.
Source: BP 2013.

Japan's wind, solar, biomass and geothermal power capacity totalled 11.3 GW by the end of 2011, with PV accounting for the largest share, followed by biomass, wind and geothermal power (PwC Japan 2012). In 2012, these 'new energy' sources accounted for less than 2 per cent of overall electricity generation (IEA 2013a). In 2011, Japan had 4.9 GW of installed solar capacity (PwC Japan 2012). Japan's installed solar PV capacity has grown significantly over the past decade and particularly since 2008 (JPEA 2013). Japan currently has the fourth-largest PV market in the world and is the third-largest producer of PV panels. With an annual increase in capacity of 200 MW, in 2011, the total installed wind capacity in Japan reached 2,536 MW with 1,832 turbines, including 25.2 MW from 14 offshore wind turbines. Most of these wind turbines are scattered around the far north and south, away from major population centres (Engler 2008, Inoue and Miyazaki 2008, Maruyama et al. 2007, WWEA 2011). In 2001, Japan only had 303 MW of installed wind capacity, so the growth over the past decade has been rapid (JREF 2013). However, the growth in installed wind capacity in Japan is low compared to 5 GW annual growths in the US and almost 20 GW in China. Italy, which shares Japan's heavy reliance on imported energy, boasts twice as much installed wind power capacity as Japan (Valentine 2011: 6849). Total energy produced from wind turbines during 2011 was 4.246 terawatt-hours (TWh) and this corresponds to less than 0.5 per cent of national electricity demand, against

13 per cent in Germany and 24 per cent in Denmark (Inoue and Miyazaki 2008). Japan's geothermal capacity stands at 2,470 MW, at 18 geothermal plants without any new projects since 1999 (DLA Piper 2012).

Institutional Bias Towards Solar Power

One of Japan's responses to the 1970s oil crises was an effort to develop alternative sources of energy. Thus, until the early 2000s solar power was favoured by the state. It was also an attempt at linking industrial and energy policies, perceiving of potential industrial profits in the energy area, drawing on traditional Japanese strengths in the manufacturing of high-technology equipment. While this was an energy security strategy first, for MITI it was also always a strategy that harboured long-term export potential. The promotion of new exports has always been at the core of MITI/METI activities. The preference for solar over wind energy has not been prevalent due to cost-effectiveness. METI (2010) estimates the cost of wind power at ¥9–14/kWh, as opposed to ¥49/kWh for solar PV. The bias in favour of solar power has been a political choice. The relative success of solar shows how much easier it is to break through if the industry is able to work within existing industrial and institutional structures, that is, when the vested interest structure actually works for it. The success of solar PV has been accomplished relatively smoothly – without changing any major institutional, industrial or organizational structures. As established in Chapter 1, it has been the same politicians, bureaucrats and business elites finding ways to pursue new industries within the bounds of the system.

Historically, there has been a policy tribe within METI supporting solar. To METI and NEDO, the success of solar gradually became a matter of prestige, largely synonymous with the success of METI/NEDO itself. This has made it much harder for opposing interests to work against solar or in favour of another renewable energy source. While solar has challenged the existing vested interest structure to a far lesser extent than wind and thus suffered far less opposition on the part of the utilities, it is unlikely that it would have risen without bureaucratic support. This it received, despite not being as cost-effective as wind. In terms of bureaucratic influence, solar is a minor player compared to the utilities. However, compared to wind, it has enjoyed a partial insider status, both because of the way it has been able to work within, rather than against, the existing vested interest structure and because of the support it enjoyed inside METI and NEDO, almost to the extent that one may suggest an institutional bias in favour of solar over wind (Bradford 2006, DeWit and Tani 2008, Kimura and Suzuki 2006, Schreurs 2002). Solar power remained the preferred renewable option for the government well into the past decade. For example, solar photovoltaic energy was accorded special treatment under the 2003 RPS program. Each kWh of solar PV electricity that was purchased equated to 2 kWh generated by other renewable energy technologies (IEA 2010a). In addition, starting in January 2009, METI provided subsidies and tax credits for the installation and renovation of solar panels on residential homes and the November 2009 FIT had been restricted to residential PV systems.

Solar PV has benefited from major government R&D, going back to MITI's 1974 Sunshine Project. Perceiving of the country's energy vulnerability, the goal was about energy as much as industry, seeking to provide a substantial amount of non-fossil energy by 2000, as well as result in industrial applications for domestic use and exports (Broadbent 2002, IEA 2008, Kimura and Suzuki 2006). In a shrewd move of bureaucratic politics, MITI realized that the bigger its request for funding, the easier it would be to gain publicity and political support (Kimura and Suzuki 2006). Thus, from 1974 the Sunshine Project received substantial funding.

The budget was increased considerably after the 1979 oil shock. This coincided with MITI losing faith in solar thermal, which had been the program's original mainstay. As MITI wanted to retain their budget, a total budget that had increased by more than 200 per cent would from 1981 onwards go primarily towards solar panels. PV thus acquired stable and abundant funding from the failure of solar thermal. A new agency, NEDO, was established in 1980 and a legal framework for fostering renewables was hammered out. The aim was that by 1990 solar power would supply 5 per cent of total energy demand and 7 per cent by 1995 (targets were never fulfilled). With MITI/METI, NEDO became the main new energy bureaucratic actor in support of solar power (DeWit and Tani 2008, Kimura and Suzuki 2006).

The program also convinced the industry that this was a promising emerging market and in parallel with the Sunshine Program several companies contributed R&D of their own, although the government's commitment was most likely the main stimulant. It created an assurance that this was a field that would receive persistent funding at a stage where no commercial profits were yet to be had. While NEC started researching solar power in the 1950s and SHARP had since 1963 dominated a minute commercial market for solar panels for lighthouses and satellites, without the Sunshine Project there would have been few incentives to get involved. However, the companies that eventually succeeded were not the giants. NEC never joined the program and Hitachi and Toshiba withdrew at an intermediate stage. Their perception in the late 1980s was that market opportunities were not developing fast enough and that solar panels would always remain a niche. Thus, they were leapfrogged by smaller companies. SHARP, Sanyo and Kyocera stayed with the program and went on to become the core of the Japanese solar panel industry. The utility companies were uninterested in contributing to a market that they considered marginal. It took a four-year (1986–1990) NEDO demonstration project to persuade them of the stability of PV. As the market for solar power kept growing, the relationship with the utilities has improved (Kimura and Suzuki 2006).

By 1993, cumulative government investments had reached ¥600 billion, with almost no commercialized technologies to show for. NEDO and MITI were under strong political pressure for the program to be commercially successful. A full subsidy and deployment program for solar seemed like the only solution

(Bradford 2006, Kimura and Suzuki 2006). This subsidy came through the 1995 *Seventy Thousand Roofs* program. It created major industry growth and rapidly falling prices, both from technological progress and economies of scale. It was also a major break with past policies, as this essentially meant a subsidy on installed residential PV systems. The JPEA lobbied heavily for this (Bradford 2006, Kimura and Suzuki 2006).

Having been supported by parts of the bureaucratic apparatus, solar has always been on the inside. And building a solar industry without MITI/METI would have been tricky. At several critical junctures the nascent industry could have fallen by the wayside. First, MITI asked for a big budget, assuming that a big project would be taken more seriously and more easily becoming a permanent budget feature (Kimura and Suzuki 2006). At the next big juncture – the lack of major progress in solar thermal – MITI shielded the budget and even received a huge increase because of the 1979 oil shock, then funnelled it into PV. As the utilities were reluctant to let solar onto the grid, NEDO's four-year pilot to demonstrate the suitability of the system forced the utilities to give in. As there has been no unbundling of electricity transmission and generation, the utility companies could have shut solar out. The JPEA, cooperating with sections of NEDO and MITI and, boosted by public pressure, successfully pushed regulatory change. When focus was cast on the lack of technologies brought to the market, MITI used the amount of money already spent to argue that Japan should fund deployment as well. Thus, solar at important junctures had major players fighting for it and committing to it, at the expense of existing actors in the Japanese energy-industrial complex (Moe 2012).

As a consequence, Japan has played a key role in the development of solar power, which has enjoyed more than 40 per cent growth annually. The installed PV power in Japan quadrupled between 2000 and 2005. Such rapid growth has helped to make Japan as the world's leading producer of PV cells during the past decade. Japan's prowess in solar technology has evolved through a combination of favourable economics and supportive government policies. Direct subsidies have been a contributing factor. The Japanese government has allocated more than $250 million per year to support solar power, significantly more than the United States or Germany (IEA 2006). With increasing economies of scale and incremental technological improvements, prices for PV installation in Japan have decreased by 7 per cent each year between 1993 and 2006. Private Japanese firms have been the primary beneficiaries of this incubator effect. In 2005, three of the five biggest solar panel manufacturers were Japanese (Broadbent 2002, IEA 2008, Kimura and Suzuki 2006, DeWit and Tani 2008). Companies like SHARP, Kyocera, Sanyo Electric and Mitsubishi Electric have dominated global cell and module production during the early 2000s.

However, the subsidy expired in 2005, partly because METI had earlier assured MoF that the subsidy would only run until self-sustained growth was achieved and partly because of a general swing in favour of market-based policies initiated by Prime Minister Junichiro Koizumi (2001–2006). The removal of subsidy shows that there are more powerful bureaucratic actors than METI and that while solar has certainly been favoured over wind it remained a minor player compared to the utilities and the nuclear lobby (Moe 2012). Since then, Japan has lost its predominance. While at the time of its demise, the subsidy had been scaled down to only 3 per cent of investment cost, or ¥20,000/kW, in 2005 it led to a 35 per cent slump in the number of applicants for residential PV subsidies, at the stage that the market took off globally. Following the Sunshine Project of the 1970s, Japan was a world leader in the solar panel industry until the early 2000s, but now lags far behind China, which was late to enter this business. Japan no longer has the world's largest installed capacity of PV and SHARP in 2007 lost its position as the world's largest producer of solar panels. Japan controls 14 per cent of the market, a significant drop from more than 50 per cent in 2004 (Roney 2010). In Ernst & Young's (2009) renewable energy attractiveness index, even in solar, Japan was ranked 11th. While Japanese firms were global leaders in the PV industry in the 1980s and 1990s (Watanabe et al. 2000), their position has eroded since then. The Japanese shares of global PV patents, solar cell production and capacity additions fell from 51 per cent, 22 per cent and 36 per cent in 1995 to 22 per cent, 13 per cent and 7 per cent, respectively, in 2009 (Peters et al. 2012).

While solar has lived a relatively charmed life until 2005, wind power has been the ultimate outsider (Moe 2012). For solar, Japan invented an incentive structure that was emulated. But its wind power industry accounts for only 1.6 per cent of the global market (*Energyboom* 2009). Ernst & Young (2009) ranks Japan 21st (of 25) in wind power attractiveness. The reason for slow penetration of wind energy in Japan is the flip side of the vested interest structure that solar has benefited from. There are no major interests speaking on its behalf and, while wind turbines ought to be a promising area of industrial success, few industries perceive of it as any natural extension of their existing activities. The vast majority of Japan's ¥100 billion renewable energy budget per year has historically been spent on PV (Maruyama et al. 2007). Even though wind would have been far more cost-effective than solar – for the entire period since the oil crises – it has never really been on the agenda. This speaks to the difficulty of transitioning to a new energy source without significant political support from powerful interests.

Solar PV and windmills supply their power in a different manner. Solar PV is typically installed on rooftops of individual buildings. Since most houses need electricity, they are already grid-connected and, consequently, the solar panel is automatically connected (at least since the NEDO demonstration project). Solar goes to the individual house, which is already a customer of the utility. A windmill is typically set up away from the grid, begging the question of who should pay for the cost of connecting it. This also means that major wind power expansion requires greater infrastructural investments than would solar (Moe 2012). Because there has been no unbundling of electricity into transmission and generation, unlike with rooftop solar power, the electric utility companies have been able to shut wind power out from 'their grid'. According to Iida (in Engler 2008), the utilities 'act as regional monopolies, functional monopolies and political monopolies.

They are rule makers and they make an effort to exclude wind power from their grid'. The ten regional utility monopolies in Japan are the gatekeepers of Japan's electricity market. The electric utilities have had enough influence in METI to get their way and understanding the factors that discourage the utilities from purchasing wind power illuminates the challenge ahead for wind power diffusion efforts (Valentine 2011: 6849). Moreover, none of the major METI tribes speaks for wind power, whereas solar has its support group. Even NEDO in the 1990s concluded that large-scale wind power was unfeasible.

For the utilities to let wind power into the grid is seen as acquiescing to a process of liberalizing the entire electricity sector, which they have fought tooth and nail for almost two decades. Wind power is produced by IPPs, which essentially means that they are competing against the utilities. They have had no interest in letting a rival power producer into the market. Thus, the utility opposition against wind has been far stronger than against solar (Moe 2012). In order to keep wind power off the grid, the utilities have imposed 'introduction limitation quotas' for wind. Wind power above these quotas has been treated as bids and selected by lottery. With bids far outstripping the quotas (sometimes by a factor of ten), the quotas have hampered expansion (Maruyama et al. 2007).

Whether genuine or merely excuses, wind power has met with a number of objections from the utilities. One is that Japanese weather conditions are particularly difficult, with choppy winds and seasonal typhoons. Between 2004 and 2007 several turbines were severely damaged, leading to new safety standards, the J-class windmill and a new building code classifying wind turbines taller than 60m as buildings. This made the application procedure complicated, expensive and lengthy. A Japanese wind power application takes from two years and upwards before a windmill is installed. The similar process in the US might take 3–4 months. This makes it hard to achieve profitability. Using wind conditions as an argument is primarily an excuse and wind conditions do not present insurmountable technical obstacles (Moe 2012).

The high cost of wind power relative to traditional energy sources, such as nuclear or coal, also helps explain why utilities have been reluctant to purchase more wind power and why the government has been hesitant to mandate higher purchases of wind power (Valentine 2011: 6849–50). Moreover, the utilities contend that the intermittency of wind power poses unacceptable risks to grid stability. The intermittency of wind is widely regarded as the biggest hurdle facing wind energy diffusion today (Ackerman 2005, Boyle 2004, DeCarolis and Keith 2006). The current technological consensus appears to be that wind energy can contribute up to 20 per cent of a large scale electricity grid's power without requiring additional backup systems to cover power fluctuations associated with wind intermittency (DeCarolis and Keith 2006). This is made possible by more effectively utilizing surplus capacity that is already built into the system (Boyle 2004). Accordingly, there are reported cases of electric utilities forcing wind energy providers to assume the cost of storing the energy generated in order to sell it to the utility in consistent flows. This requirement increases the cost of wind energy by as much as 50 per cent, severely curtails profit margins for wind energy providers and dampens market development (Valentine 2011: 6850).

Even if the cost gap between wind energy and Japan's dominant electricity sources could be narrowed, there is a prevailing sense amongst members of Japan's wind energy community that utilities resist wind energy because the technology is an operational bother. The aforementioned intermittency challenge exemplifies an operational drawback. Although higher amounts of wind power can be incorporated into existing electricity grids without necessitating increases to reserve capacity, the inherent power fluctuations complicate the dynamics of electricity supply planning and require system adjustments. Without incentives, incorporating more wind capacity is an added inconvenience that monopolies can do without. Another operational bother is the added work involved in integrating a plethora of wind turbines into the electric grid. If a utility wishes to add 1 GW of generating capacity by constructing a nuclear power plant, grid planners would have at least three years lead time to plan a grid connection to the site. Conversely, to generate an equivalent amount of power (inclusive of load factor differences) through wind energy, over 1,000 2-MW turbines would be required. Grid planners would have to coordinate grid connections to a number of sites within construction lead time intervals that can be as short as 4–6 months (Wizelius 2007).

Finally, the utilities oppose wind energy as it threatens to upset their monopolized business model. Incorporating wind energy into the electricity grid poses a dilemma for utility strategists. On the one hand, a utility could decide to develop wind energy projects itself; thereby, incurring the time-consuming obligations associated with site selection, project planning, community relations, environmental impact assessments, project management, etc. On the other hand, the utility could decide to avoid the logistical bother and purchase wind energy from IPPs. However, delegating responsibilities for generation will result in profits and control leaking from the monopolized supply chain. Moreover, under either option, lower margin wind energy would displace profitable energy generated by conventional technologies. Clearly, neither of the alternatives holds much appeal (Valentine 2011: 6850–51).

Consequently, the major reason why wind has not flourished in Japan is that it challenges the existing institutional framework to a much greater extent than solar (Moe 2012). In terms of industrial policy, solar was a more obvious fit. The potential customers of solar PV were individual households (and businesses) and it was easier to create discreet products sold to individual customers. From the inception of the Sunshine Project, the amounts of conceivable products arising from solar technologies were considered far more numerous than from wind. MITI always saw a greater potential for commercialization from solar and, unlike wind, solar fit into MITI's strong preference for high-tech export industries. Even in recent years, METI's solution to solving the economic crisis has been to advocate exports rather than structural change (MoFA 2008). Consequently, if a distinction were to be made between energy and industrial policy, wind power in Japan belongs almost exclusively to the former. It is about increasing energy supply. Solar policy on the other hand was always energy and industrial policy (Moe 2012).

Structural economic change, like the rise of renewables, typically leads to resistance from the existing vested interest structure, which feels it stands to lose from the rise of new industries. According to Moe (2012: 260), in the Japanese case, there are major differences with respect to how the state has dealt with solar and wind energy. In fact, the solar industry has been far more on the inside of this structure than wind industry. The Japanese vested interest structure, which consists of the METI, as the energy policy-making hub, working closely with and favouring the interests of the electric utilities and the nuclear industry, has persistently kept wind power on the periphery, showing how hard it is for an industry challenging the existing structure to rise. Of the renewables, solar has been the insider, to some extent even benefiting from the structure. It needed government support, but did not require major structural change. Rather, it rose within the industrial and institutional framework, drawing on expertise within engineering and precision mass manufacturing in existing Japanese manufacturers. Instead of challenging the structure, solar complemented it, with solar panels becoming yet another successful Japanese export industry. Still, compared to traditional energy actors, solar is a minor player (Moe 2012: 260).

The Feed-in-tariff and the Local Level Push Towards Renewable Energy

The new FIT has been in effect since July 2012, encompassing PV, wind power, small hydroelectric, geothermal and biomass (Ayoub and Yuji 2012, IEEJ 2011c: 5). It obligates Japanese electric utilities to purchase electricity generated from renewable energy sources, including solar PV, wind power, hydropower, geothermal and biomass at fixed FIT prices, which are higher than normal contractual prices and applicable for a fixed period of 10 years for residential PV power and for 10 to 20 years in the case of commercial PV power or other renewable energy power generators. The new FIT substantially expands the scope in terms of energy sources to include all renewable energy and designates all of generated electricity as purchasable (IEEJ 2011c: 6). The extra costs incurred by electric utilities in the purchase of renewable electricity will be recouped through a surcharge added on top of the normal electricity bill to end users. Concerning the pricing, the existing rates of ¥42/kWh have been applied for sellers of excess power from residential PV systems only and ¥34/kWh for double generation users with a combination of fuel cells and PV. Purchase prices of ¥23–35/kWh have been applied on other types of renewable electricity, with duration between 10 and 20 years (METI 2012c: 5).

This type of scheme has been considered as one of the most positive measures to accelerate introduction of renewable energy (IEEJ 2011c: 5). It has been adopted in over 80 countries and states, supports roughly 75 per cent of global solar and 45 per cent of global wind and has been deemed by the World Bank and the IEA as the most effective policy tool for diffusing renewable energy and reducing its costs. The FIT plays a critical role in the energy revolutions underway in Germany, Spain, China and elsewhere. A comprehensive FIT could increase electricity production from new sources by 40–50 billion kWh or more (roughly 4 to 5 per cent of Japan's current output) in ten years, or approximately half of the increase desired by 2030 (Duffield and Woodall 2011: 3747).

The FIT does two important things. First, it guarantees a long-term market for power produced by renewable sources. Second, it pays producers a premium price to reflect the currently higher cost of generating that power relative to conventional power. That guaranteed price, of course, represents a cost to the utility, which is compelled by law to purchase the power from a legitimate renewable power producer. However, the FIT allows the utility to pass that cost onto the customer, meaning households and businesses that consume that electrical power. This allows the cost of shifting to renewable energy sources to be spread out thinly over households and industrial sectors. This is crucial during the early stages of renewable production while new technologies and a larger market share make possible sharp cost reductions (DeWit 2011).

The new FIT represents a policy opportunity for diffusing alternative power as rapidly as possible, thereby contributing to the resolution of Japan's power needs as well as strengthening its position in the global competition to promote green businesses. A green energy policy at the core of a smart post-Fukushima reconstruction could create robust and sustainable demand in Japan's domestic economy, opening up lucrative opportunities and generating both electricity and jobs. To grow renewable, distributed power requires smart energy policy, centred on a robust FIT, as well as a strong dose of deregulation and competition for the monopolized utilities (DeWit et al. 2012: 168–69).

While the new FIT is a major step forward towards increasing the share of renewable energy in Japan's electricity portfolio, interests of electricity monopolies may still affect its successful implementation. The nuclear village views this policy opportunity as a threat, as it is aimed at cutting dependence on nuclear energy (DeWit et al. 2012: 166) as Japan is handicapped by the opposition of vested interests to changes in the status quo. The LDP, *Denjiren* and *Keidanren* went against the new FIT, resulting in a compromise whereby energy-intensive industry is given an 80 per cent discount on any increase to their electric bill due to the FIT (DeWit and Kaneko 2011). Moreover, electric utilities can refuse to purchase the power from renewables if 'it is deemed to impede undisturbed supply of electricity' (METI 2011d). Preventing unwilling utilities from exploiting this loophole will be extremely difficult (Huenteler et al. 2012: 9). According to IEEJ (2011c: 6), renewable energy power generation capacities in Hokkaido and Tohoku areas, where there is large potential for wind and other resources, have already hit the ceiling that regional electric utilities have set for undisturbed supply of electricity, rendering the Act ineffective in accelerating introduction of renewable energies. While the FIT has led to major new optimism in the renewable industry (although more so for solar than wind), it is still too early to predict its long-term effect (Moe 2012). The challenge for Japan is that FIT will result in higher electricity prices (approximately ¥2.73/kWh by 2020) and this may bring the possibility of suspension or revision to its FIT in order to control the rise in electricity tariffs caused by the growth in installed capacity of renewable energy (Interview with Hiroshi Hamasaki, 16 January 2013).

Aside from introduction of a new and improved FIT, since the Fukushima disaster, the momentum for moving into renewable energy has been striking at the local level in Japan. Soon after the disaster, the nuclear village became the focus of an increasingly powerful challenge from advocates of renewable energy. The most prominent actor in this expanding campaign is Softbank CEO Son Masayoshi and his fast success in organizing local governments. His Renewable Energy Council for prefectures has enrolled 35 out of 47 prefectures and a similar council for 'designated cities' with a population over 500,000 includes 17 out of 19 such cities. Son was also instrumental in keeping FIT on the policy agenda during the fraught months following the Fukushima disaster (DeWit 2011, 2012a, Wakamatsu 2011).

Son's goal is to encourage local governments to shift to renewable energy sources and to provide land for mega-solar panels, wind turbines, geothermal, small hydropower generators and other renewable energy forms. Investment would come from Son's fund as well as other public and private sources. The key to success is participation by prefectures, which would share the financial benefits of having solar and wind farms and other natural energy sources located on their land. But the challenge is the scarcity of available and viable land. Son hopes to raise the amount of electricity generated by renewables to 20 per cent by 2020. This would include 100 GW of solar energy and 50 GW of other renewable energies. Geothermal, in addition to solar and wind, Son believes, has great potential for development, given that Japan has geothermal power sources equivalent to twenty nuclear plants (Johnston 2011).

Local governments have been particularly aggressive in responding to the crises driven and exacerbated by the Fukushima shock. The effective collapse of national energy policy has seen many rethink their growth strategies and revamp their intergovernmental organizations, both among themselves as well as between them and the central government. The Fukushima shock was profound for most local governments due to the existential threat to power supplies as well as the central government's abysmal crisis management in the weeks following the disaster (DeWit 2012b).

Major local governments such as Metropolitan Tokyo and Osaka are especially concerned by their vulnerability to highly centralized power generation and transmission as well as its clearly incompetent governance by the national administration. One of their responses to this threat from centralized, overly complex energy institutions dominated by vested interests has been to increase local resilience and autonomy via decentralized power generation. Tokyo, for example, determined that it needed its own generation capacity in order to maintain subway transport and other critical functions in the event of an emergency. Consequently, it is installing gas-fired power and a small-scale smart grid separate from the

TEPCO utility. Moreover, Osaka City and Osaka Prefecture have banded together to launch an energy commission. They are explicitly committed to ramping up conservation and renewables in the face of the central government's inertia. Kobe and Kyoto have joined Osaka as partners in the effort. Other prefectures, including Kanagawa and Saitama, are also explicitly aiming their policy-making at efficiency and fostering an energy shift to renewable power so as to enhance self-reliance, employment and business opportunities, as well as international competitiveness. While not tallied yet, the locals' investments in conservation and energy efficiency are many multiples of the budgets for renewables. The central government's FIT adds to these kinds of generalized incentives to enhance local resilience (DeWit 2012b).

Renewable Energy Potential and Structural Constraints

Huenteler et al. (2012) argue that the current regulatory environment, with the new 2011 law, which extended FIT, creates a situation similar to that under the *German Renewable Energy Act*. Indeed, the 20 per cent target formulated by former Prime Minister Kan implies an electricity sector transformation similar to the one that took place in Germany in the last decade. Germany produced 7 per cent of its electricity from renewables in 2000; while renewable sources accounted for 8 per cent of Japan's electricity production in 2009. In the decade after 2000, Germany installed some 43 GW of renewables and increased their share to 17 per cent of electricity production. Japan's aim to raise the renewable share of power generation to 20 per cent by 2020 would require roughly 70 GW of capacity (Huenteler et al. 2012: 7). Since Japan's hydroelectric potential is largely exploited (Zhang et al. 2012b), virtually all of the additional renewable generating capacity called for in the 2010 BEP would have to come from other sources.

Most of the increase in renewable energy is expected to come from solar energy by encouraging the installation of solar panels on roofs and developing larger-scale solar facilities. As discussed above, having long been a 'pet project' of the METI, PV seems perfectly positioned to play a major role in the proposed transition (DeWit 2009). This is not only because PV plants are quick to install, but also because they are suitable to fill the current gap between electric capacity and peak demand around midday. In June 2011, the government announced a goal of installing PV systems on ten million roofs by 2030. Also, the revised FIT envisages PV to account for more than 80 per cent of newly installed capacity in the coming decade (METI 2011c). The new FIT, too, lists promotion of the domestic industry and thereby strengthening the international competitiveness of Japan as one of its main targets. Arguably, although no longer global leaders, Japanese industrial expertise in solar energy is significant and, among the political proponents of renewable energy, solar energy enjoys a special position (e.g. calls for a 'solar belt' in East Japan) (Son and DeWit 2011). Japan is still a net exporter of PV cells, and domestic firms may regain momentum from the new FIT. The incremental costs of installed capacity under the FIT will be lower than a decade ago due to rapid reduction in price of solar cells and Japan's relatively high electricity prices (Huenteler et al. 2012: 9).

Green energy investment jumped 75 per cent to $16bn after introducing green subsidies following the 2011 Fukushima nuclear disaster. Many of Japan's largest corporations, from steel mills and automobile makers to ceramics and electronics makers, are developing renewable technologies, often incorporating solar and wind power features into their offices and factories. Following the start of the new FIT system for electricity from renewable energy in July 2012, many large-scale solar power generation projects have been launched. The FIT system has made a promising start mainly for solar power, with an 80 per cent share of new projects until November 2012. The installed capacity of solar power is increasing rapidly, driven by the FIT which was launched in July 2012. In the first three months since its enforcement, solar power generation plants worth 1.5 GW have been licensed for the FIT system, a large increase from the cumulative capacity of 4.9 GW at the end of 2011. Between April and December 2012, Japan installed 1,178 MW of renewable energy capacity, including 1.12 GW of PV and newly licensed capacity between July and November 2012 to qualify for the FIT system reached 3.64 GW. By the end of 2012, IEEJ (2013e: 6) estimate that Japan has installed solar PV capacity of 7 GW. The high pace of introduction continued above expectations by mid-2013 (with 12 GW of additional capacity licensed since July 2012) as profitability of renewable generation projects was maintained due to the generous purchase price of renewable electricity in the first three years of the FIT system (IEEJ 2013a: 12; IEEJ 2013e: 6).

For example, Japan is set to build its largest solar power plant, costing ¥27 billion, in Kagoshima City. This plant, expected to be completed in late 2013, will supply 78,000 MWh of electricity every year and is likely to fuel approximately 4,333 households, based on their average consumption of 18,000 kW a year. New 'Mega Solar' power plants, such as the one in Kagoshima, are being built on land that has been idle for a long time, land that was purchased and prepared expecting business growth, but became unnecessary as the economy slowed down and has since remained idle. Having already been prepared for business usage with electricity transmission lines nearby, such land is ideal for solar power. However, the amount of such land is limited and will be used up in the near future (IEEJ 2012h: 6).

Technical wind potential in Japan is significantly higher than current installed wind power capacity (Valentine 2011). Data compiled by the Geographical Survey Institute in Japan indicate that a mid-range estimate of 70,000 turbines could be situated at 964 prospective onshore and offshore sites (Ushiyama 1999). Assuming a turbine power rating of 2 MW, this implies that there is 140 GW of wind power potential in Japan, almost 75 times current installed capacity. In the short run, NEDO foresees at least 10 GW of installed wind capacity as being a feasible target by 2020 (Inoue and Miyazaki 2008). In short, there is evidence of significant amounts of untapped wind power potential in Japan.

Following the Fukushima disaster, full-scale experimental research on floating wind turbines has started in Japan. In March 2012, a consortium of Japanese companies including Marubeni and the University of Tokyo was awarded a contract for experimental research by the METI. In the first phase of the project, a 2 MW floating wind turbine and a floating transformer station will be installed off the coast of Fukushima prefecture in 2012. Two 7 MW wind turbines will then be added in the second phase from 2013 to 2015. The research aims to establish a business model for floating offshore wind power generation and to develop it into one of Japan's major export industries (IEEJ 2012c: 4).

Research suggests that up to 20 per cent of Japan's electricity could be generated through intermittent renewable sources (i.e. solar PV, wind, wave or tidal power) without requiring additional storage or generator backup. Even under the current electricity costing system (in which wind energy is 50 per cent more expensive than conventional energy sources because externalities are not internalized), a 20 per cent contribution of wind power would have a low impact on aggregate energy prices and even a lower impact on corporate profitability. A 20 per cent contribution from wind would result in an aggregate electricity cost increase of about 10 per cent. In firms where energy costs represent 10 per cent of overall operating costs, this increase would amount to a 1 per cent increase in operating costs (Valentine 2011: 6851).

Despite the potential for an increase in the uptake of renewable energy in Japan, sceptics point to various structural issues that may impede the pace and scale of this uptake. Scalise (2012b) argues that various issues may prevent a relatively fast uptake of renewable energy in Japan and that Germany's achievements with renewable energy do not imply that Japan can replicate them. Virtually all OECD countries provide 'new renewable' generation at some level. However, geography, market structure and government policy determine the quantity. Denmark (wind), Germany (solar), Iceland (geothermal) and Spain (wind/solar) provide 15–30 per cent of their total generated electricity from new renewables. In contrast with Japan, however, all four leading countries in renewable energy had relatively low electricity prices in 1990 before the introduction of feed-in-tariffs. In addition, electricity capacity reserve margin (the percentage of installed capacity in excess of peak demand) – a common metric for surplus capacity – indicated percentages well above 20 per cent. This pre-existing oversupply prevented the intermittent supplies generated by new renewable from risking blackouts as surplus back-up power existed in the event that solar, wind and other renewable were unable to meet peak demand (Scalise 2012a: 147).

With Japan's national reserve margin in the low 10 per cent range and falling year-on-year (Scalise 2011b, Scalise 2011c), rolling blackout risk places renewed emphasis on rapid investment from stable sources with relatively quick lead times in the siting, licensing and construction of new generation capacity. This economic reality militates against strong support for new renewable among industry and the incumbent suppliers in the short- to medium-term (Scalise 2012a: 147). Japan also has higher electricity rates than most other advanced countries and much of this is due to the rising costs of finding sites to locate electric plants in the face of local opposition commonly known as not-in-my-backyard (NIMBY). This presents a problem for renewable energy, such as wind and solar power because they require far more land than nuclear or fossil fuels to produce a kWh of electricity (Scalise 2012b: 8).

The difference between Japan and Germany is that Japan is not connected via long transmission wires to cheap electric power in South Korea, Russia or Taiwan. By contrast, Germany is connected to the rest of Europe through an international grid, which helps them in the buying and selling of surpluses and deficits of electric power. Because electricity cannot be stored, countries relying on volatile sources like wind and solar need to have a safety net of being able to import when they run into a deficit situation (Scalise 2012c: 8). According to Scalise (2012c), another difference between Japan and Germany, as well as other countries which have chosen to move away from nuclear power, is the high population density ratio in Japan, i.e. the number of people per square km. The more people that are located in a certain area, the harder it is to site, license and construct plants. Consequently, the cost of NIMBY is likely to increase the construction costs for all power sources moving forward, given Japan's high population density ratio and size of its economy. Lead times grow as a result of NIMBY. As it becomes increasingly difficult to find site locations which are both technically and politically feasible over time, it becomes increasingly expensive to secure support from local population. Not everybody is willing to host a facility unless the price is right. And as witnessed with siting issues related to nuclear power plants, these negotiations can go on for many years, if not decades.

Putting aside the issue of financial costs, i.e. ¥/kWh, according to Scalise (2012c), a far more pressing issue is that of energy density, or watts per square meter. If Japan abandons a high-density baseload power source, such as nuclear, which on average has a density of 1,000 W/m^2, what renewable options can replace it? Solar, especially large-scale PV power farms, has a power density between 3 and 10 W/m^2 – no more than 1 per cent of the density of nuclear power. Power density of wind in Japan – where the wind is slower than in much of Europe – is even worse at just 1.3 W/m^2, about 0.1 per cent of nuclear's yield. Biomass has a very low power density of 0.3 W/m^2 (Scalise 2012c: 8–9).

If Japan were to replace a 1 GW of nuclear power plant with a large-scale solar farm, it would need 444 km^2 of land. In the United States, solar farms are largely located in the desert and there are no deserts in Japan. Of all of the land in Japan, 65 per cent is designated as forest. The forests are not suitable for siting solar farms because of environmental issues. Another 11 per cent is farmland and the remaining 24 per cent is mostly urban or suburban. Currently, there are about 158 sites around Japan marked out for solar power, with a total installed capacity potential of 151 MW. Most of these sites can yield less than one MW and they are largely on rooftops in urban areas. For wind farms, there are currently 372 sites, 110 of which have a potential capacity of less than 1 MW and only 91 have a potential or more than 10 MW. Altogether these 372 wind sites could produce 2.5 GW of electricity, only about 0.9 per cent of Japan's current electricity capacity of 280 GW (Scalise 2012c: 9). Consequently, given these structural constraints, Scalise (2012c) argues that the notion of supplying 20 per cent of electricity from renewables by 2030 is over-optimistic, given our knowledge regarding the watts per square meter requirement, not to mention the economic costs and the environmental costs involved (Scalise 2012c: 9).

In contrast to solar power, the cost of wind power in Japan has been largely economically viable. Wind power is far less costly per kWh than solar and offers a slightly higher utilization rate. Japan has a large wind power potential, most of which lies in the Tohoku and Hokkaido regions. Despite the large potential, Japan currently lags behind many other major wind power producing countries due in particular to the limited progress made in the field in the past few years and due to political issues that prevented the increase of wind power, as discussed above. The main limiting factors for wind power deployment in Japan include grid stability, mountainous geographical conditions and various environmental restrictions, particularly the protection of Golden Eagles. For example, its unreliability requires increasing fossil fuel back-up sources while the windmills pose a danger to avian wildlife (Scalise 2012a: 145). Competition for land use force wind sites to more remote, less urbanized areas which gives rise to increased transmission costs (Wizelius 2007). The capacity for Japan to significantly expand its wind power is also limited by a lack of space and frequent hurricanes, which can damage wind turbines (Meltzer 2011).

In addition, the most productive sites for wind power are located far from where the electricity is needed, necessitating the construction of new power lines often in the face of local resistance. Quantitatively, one wind power unit (a windmill) provides far more energy to the grid than one rooftop solar panel. Thus, wind provides more power per energy unit and requires a bigger investment in terms of grid lines. But the grid is weaker in the country, where most windmills are located. With solar, each power generator is too small to affect the grid by much. Also, they are typically located in populated areas, where the grid is already strong. For Japanese wind the most important issue is priority access to the grid. Without this, wind power cannot grow, almost regardless of other regulatory changes, such as the introduction of new FIT. The degree of structural change required is far greater than for solar (Moe 2012). These factors have adversely affected the wind power sector's development plans and are likely to remain as potential bottlenecks in the future (Bhattacharya et al. 2012: 20). Given a wide range of constraints to faster penetration of wind power, it is not surprising that until November 2012, wind power accounted for only 14 per cent of approved new renewable energy projects since the introduction of the new FIT.

Offshore wind power potential in Japan is high but has yet to be aggressively exploited due to the higher costs associated with offshore wind power development. A government feasibility study is expected to yield a pilot offshore wind turbine within a few years (Engler 2008, Moe 2012). TEPCO is involved with the University of Tokyo to develop a floating wind-farm. Developing offshore wind energy is challenging for Japan, as the country has limited experience in establishing offshore wind turbine technologies as opposed to international rivals. This is partly due to the fact that Japan must use the floating type of turbines to utilize offshore wind resources since the deep seas surrounding Japan limit the installation of bottom-fixed systems. In fact, at the test sites 20 to 40 km offshore, the water is as deep as 100m to 150m. Another reason is that Japanese turbine makers are far behind major competitors in the land based turbine market. Japanese turbine manufacturers have only a 3 per cent share of the world market and only 20 per cent even in the domestic market (IEEJ 2012c: 4). Regardless of technological issues, offshore wind farms are likely to draw the ire of fishermen and environmentalists as they pose a danger to fish habitats and avian wildlife. Iida (in Engler 2008) warned that offshore windmills might face vested interest problems from the fishing industry, which is one of the foremost clients of the Ministry of Agriculture, Forestry and Fisheries.

Despite the fact that its abundant geothermal resources are ranked third in the world, the geothermal generating capacity in Japan has remained low. Geological estimates show that ample exploration opportunities exist. Geothermal power would appear to be an attractive option for Japan given that there are more than 100 active volcanoes and thousands of hot springs. Geothermal is also considered to be already economically competitive with conventional fossil fuel-fired technologies. Adequate technology and experience are available, as Japan has supplied 70 per cent of the geothermal plants worldwide. Nevertheless, geothermal power has been stalled in Japan mainly due to its high generation costs and regulations on project developments inside national parks (since 1972), where more than 80 per cent of Japan's high-temperature hot water resources are located (IEEJ 2011b: 7). The main constraint for geothermal power is that many of the promising heat sources are in environmentally sensitive areas such as nature reserves, where installation is prohibited (Bhattacharya et al. 2012: 20). Some of the best locations are in national parks, which have strict limits on their development and hot springs, which are attractive for tourists. The Japanese *onsen* – spas that rely on underground hot water – have been opposed to the development of geothermal energy because of concerns it will reduce the availability of hot water. Consequently, the ability for Japan to develop this energy source is constrained, as geothermal power faces political opposition from environmental activists and small business owners alike, who disapprove of unpredictable exploration prospects in environmentally fragile locations (sometimes national parks), where, for example, it is hot enough to produce geothermal steam close to the surface (Scalise 2012a: 145).

The challenge for Japan is to relax the regulation regarding national parks to enable installation of geothermal power plants. If the Japanese government eases the regulation on developing geothermal projects in national parks, a significant room will be opened with improved economics and many Japanese companies have responded to the potential policy change with various projects to access geothermal energy resources. The Japan Geothermal Developers' Council (JGDC) estimate that a 620 MW of geothermal power could be introduced at an electricity purchase price of ¥24.5/kWh, which could increase by another 1,670 MW if new projects are allowed within the currently prohibited national parks. The MoE estimate geothermal introduction potential at 4,230 MW, while METI puts the figure at 1,130 MW, both with a power purchase price of ¥20/kWh. However, JGDC, an industry association, indicates a much lower estimate at approximately 300 MW, saying that projects would become hardly commercial at a purchase price below ¥20/kWh (IEEJ 2011b: 7–8). While the new FIT has set the purchasing price for geothermal power at ¥27.3/kWh, given that the government has not eased the regulation on developing new geothermal projects in national parks, no new geothermal projects have been approved since the FIT was passed in July 2012.

The potential for renewable energy – mainly including PV and wind power – is limited in Japan due to physical–geographic reasons and constraints in technology and system integration (Zhang et al. 2012b: 987). Solar power is quiet and clean, but high cost per kWh and low utilization rate ensure its marginalization for energy-intensive industries requiring stable baseloads to operate efficiently during business hours (Scalise 2012a: 145). Given the intermittent nature of many renewables, the amount of capacity that must be built to produce every kWh of electricity will be several times greater than for other sources, greatly reducing their cost-effectiveness. According to one estimate, even 100 GW of installed PV capacity, or the equivalent of nearly 40 per cent of current power generating capacity, would meet only 12 per cent of Japan's electricity demand. For all these reasons, METI predicted in 2009 that the share of the primary energy supply provided by renewables in 2030 would reach only 11.6 per cent, even with 'maximum introduction of technology' (Duffield and Woodall 2011). Moreover, various groups oppose renewable energy and, specifically geothermal and wind energy, which is likely to impede their future growth. Among the influential opponents of renewable energy are the fishing industry, opposing offshore renewables and the *onsen* industry, opposing the exploitation of geothermal energy (Vivoda 2012, Huenteler et al. 2012).

Given numerous structural and practical impediments, it will be difficult for Japan to rely on renewables to replace nuclear power. The new FIT law will undoubtedly expand the share of renewables from the current level but, given numerous constraints, it is unlikely that renewable energy will account for 20 per cent of Japan's electricity supply by 2020. While Fukushima is likely to contribute to more rapid deployment of renewables, the main problem may be transmission and distribution. Arguably, closing the new FIT's grid stability loophole must be the first priority (Huenteler et al. 2012: 9). Moreover, according to Duffield and Woodall (2011), another obstacle in the path to greater reliance on solar power (and other renewables) is that the existing power system could accommodate enough PV generating capacity to provide only about 6 to 8 per cent of the electricity supply. Improving the grid will require major investment and protracted fights with local governments. The IEA (2011b) suggest that Japan has the potential to increase the share of renewable energy of its electricity supply up to 19 per cent with construction of new transmission lines.

Unless there are significant technological breakthroughs in renewable energy, it is reasonable to expect that most alternatives to nuclear power will be restricted to electricity from fossil fuels (Hayashi and Hughes 2013). Renewable sources start from a very low base and if other limitations to renewable energy are considered the idea that renewable energy can replace nuclear energy could not be further from reality due to constraints on supply capacity and supply intermittency (Miyamoto et al. 2012: 34). A respondent from METI pointed out that it is too optimistic to suggest that the share of renewable energy will grow to 20 per cent by 2020, citing structural issues that were outlined above (Interview with Shinichi Kihara, 17 January 2013). Moreover, a fundamental move away from nuclear power towards renewable energy sources would require more than just technical blueprints and economic incentives to surpass Japan's structural challenges; it would require a shift in actor perceptions that form the 'iron triangle', which has historically been largely opposed to the greater access of renewable energy. Consequently, the government will have to work towards reducing the impact of industry interests on the regulatory process if any policy for renewables is to be effective (Kanie 2011a).

Conclusion

Japan played a major global role in the development of solar energy from the late 1990s until several years ago. What has been lacking, however, is any substantial policy support for raising the share of renewables in Japan's energy mix. However, following the Fukushima disaster, the improved and generous FIT has been adopted in order to increase the share of renewable energy in Japan's electricity supply. Consequently, the renewed interest in the alternative energy sector in Japan since the end of last decade and the need for more alternative sources in the aftermath of the Fukushima incident have resulted in renewable energy emerging as an alternative to fossil fuels and nuclear power. Although the level of renewable electricity deployment largely depends on the FIT levels, expectations for large-scale renewable energy deployment in the coming decades are high. This chapter demonstrates that even if the high cost issue can be overcome, a number of major challenges stand on the way to a greater penetration of renewable energy in Japan. One of the major issues is whether Japan can overcome the decades-old institutional bias towards solar power. Moreover, numerous structural issues pertaining to various renewable energy sources may stifle the potential for increased penetration of renewable energy, notwithstanding the generous new FIT and increased grassroots level support for renewable energy post-Fukushima.

Chapter 8
Electricity

Introduction

Most of the challenges discussed in the previous five chapters affect Japan's electricity supply. This is particularly the case with regards to nuclear power and renewable energy. This chapter explains how these challenges interact with the unique structure of Japan's electricity market, dominated by ten regional electric utility monopolies. Revisions of the *Electric Utility Industry Law* between 1995 and 1999 permitted new entrants, weakened central control of pricing and allowed more competition on the retail side. However, these piecemeal reforms have not resulted in lower electricity prices and have not challenged regional dominance of electric utilities. The utilities remain in control of both electricity generation and transmission, which presents an enormous obstacle to potential new entrants. At the same time, the electric utilities have been affected by the nuclear power shutdown and have found themselves in a difficult financial position as they are bearing the increased cost of energy imports to fuel their thermal power plants. The utilities, as part of the vested interest structure, remain committed to nuclear power and are also largely opposed to an increased penetration of renewable energy due to high costs. Finally, Japan's electricity grid is fragmented and there is limited capacity to transfer electricity between regions and between eastern and western Japan, which was a major issue in the aftermath of the Fukushima disaster. Consequently, with the discussion centred on the role of electric utilities, this chapter evaluates major challenges in Japan's electricity market. The chapter begins with a brief survey of historical evolution of Japan's electricity market in order to map the emergence of ten regional monopolies as key players. Before turning to analyses of the major challenges, the chapter then outlines electricity sector organization and supply structure.

Historical Evolution of Japan's Electricity Market

After a brief period of government wartime control, nine private power companies were reorganized in 1951 (ten since 1972) to assume control of most of the generation, transmission and distribution businesses of electric power throughout the country (Samuels 1987: 158). They were given regional monopolies to generate, transmit and distribute electricity within their assigned territories. This reclaimed monopoly status was the result of successful lobbying by utilities to return to a system of centralized private power, reminiscent of the early 1930s, not of the highly competitive era of the 1920s. The greatest strength of the new '1951 system' was that it exploited economies of scale, maintained stable electricity prices for over two decades and kept SO_x and NO_x emissions in check. Its greatest weakness was that it eventually disproved the theory; the 1951 regulatory structure failed to reconcile rising demand with mounting variable (fuel) and fixed (capital) costs. Over time, the weaknesses became more pronounced. From 1975 to 1985, average monopoly tariffs increased by almost 100 per cent from ¥11/kWh to ¥21/kWh. Utility companies argued that the phenomenon was the result of exogenous variable shocks stemming from an overdependence on imported fossil fuels. Although true to an extent, average tariff rates and the unit cost of fuel per kWh continued to decouple after 1985. The lack of continuity became more pronounced as market rigidities and rising capital costs prevented a nominal readjustment to pre-1973 tariff levels. Nevertheless, the original presence of 'natural monopoly' technologies continued to provide justification for a regulated power industry and accordingly to impose entry barriers, price controls and obligations to serve.

From 1951 to 1973 the electric utilities operated fairly independently as the industry transformed itself from coal to oil. They often clashed with MITI in their commercially driven haste to abandon coal for oil. But the oil shocks pushed the utilities closer to MITI. MITI helped them to operate a 'convoy' system, co-ordinating price movements across the industry and playing the role of central dominant actor in the process of reactor site selection and instigation. In return, MITI expected the utilities to play a central role in regional economic development and utility chairmen were invariably chairmen of the relevant regional economic association (except for Tokyo). Unlike other industrialized nations, the post-war regulatory regime did not break down with the oil crises of the 1970s. Rather, MITI overcame the crises through close cooperation with incumbent utilities, preserving the vertically integrated monopoly structure. The exploitation of economies of scale while controlling market power was not only considered the progressive 'public interest' view of the country's utility regulation until the early-1990s, it was also a means of preserving the political status quo (Scalise 2004: 163–64).

Following the oil crises of the 1970s, utilities became the pivotal force in government energy-diversification policy. The regional utilities played a pivotal and unique role after the first oil crisis. The utilities were prime candidates in 1975 to lead the effort for fuel substitution and diversification, for several reasons: the dominance of oil in electricity generation (74 per cent of the total electric-power output of the nine regional utilities was generated using this fuel); their prevailing aim of securing supplies for their customers; and their relatively easy access to capital. The technologies for fuel substitution and their commercial consequences in electricity production were already known, because of previous experience with nuclear and coal-fired generation. The scope for substitution was large, as more than 20 per cent of the nation's oil supply in 1973 was used by the electricity-generation industry (Nemetz et al. 1984/1985: 566).

Until the 1990s, the power utilities degraded themselves to spiritless government agency-like entities. Inter-company competition more or less disappeared and international price competitiveness declined sharply, as MITI routinely allowed the companies to pass on rising fuel costs to consumers, notably to residential consumers who paid three or four times more per kW than US consumers (Samuels 1989). The utilities were often seen at this time as quasi-government agencies. But it is hard to say who really ruled. MITI did not direct the companies and, according to Samuels, the companies were often more the 'principal architects' of MITI's plans than the 'victims' of them (Samuels 1989: 636). The companies were content to proceed with ambitious promotion of nuclear power – as long as MITI paid the bills and gave them favourable tax treatment on their nuclear accounts. But where there were disagreements on the relative roles of coal, oil and nuclear or over electricity rates, these matters were generally resolved congenially (and usually in favour of the utilities) behind closed doors. The utilities co-ordinated with each other to set rates and MITI generally approved their proposals. As Samuels (1987: 234) describes it, the utilities have 'separated state aid from state control' to a significant degree. The state has socialized risks while allowing the companies to operate with commercial freedom. Although the internal history is obscure, it is apparent that the utilities 'captured' many of the institutions that were supposed to oversee or regulate their domestic electric power operations (notably the STA and JAEC).

From the mid-1990s, the relationship became somewhat more complex. In an era of cheap oil, economic stagnation and external pressures to free up Japanese domestic markets, government emphasis shifted to the liberalization of domestic energy markets to drive down prices and the pursuit of the reduction of GHG mandated by the Kyoto Protocol. This liberalization began to, at least on paper, create more space between the utilities and METI. Revisions of the *Electric Utility Industry Law* (1965) between 1995 and 1999 permitted new entrants, weakened central control of pricing and allowed more competition on the retail side (also allowing the electric utilities to engage in other businesses, such as gas). At the same time, however, in terms of nuclear power, the utilities worked more closely than ever with METI to push nuclear power forward and silence popular opposition, as illustrated in Chapter 6.

Industry restructuring began in earnest with the advent of the Hosokawa Morihiro coalition government in 1993. By that time, electricity prices from 1983 to 1993 were not only perceived to be high domestically, but also internationally. Pre-tax Japanese electricity tariffs were three times the average US price in 1993. Broadly speaking, pressure to enact reforms was initiated by several actors, including the incumbent utilities concerned with the threat of self-generators capturing further market share; the export-led industries burdened with an appreciating currency and mounting operational costs; the potential new entrants seeking to duplicate their success in foreign deregulated markets; and the central government, which was looking not only to maintain its legal powers over the market, but also to placate foreign government pressures (Scalise 2004).

In response to such growing pressure to lower electricity tariffs, the government decided to undergo regulatory reform. Revisions to the *Electric Utility Industry Law* were intended to produce a more competitive market, albeit incrementally. Phase one (1995) permitted the nine major power companies (excluding Okinawa) to act as 'single buyers' through a revision of the Law. A bidding structure organized independently of government control was established whereby IPPs and other self-generators could bid for new contracts for their supplementary power needs. Major players in phase one were predominantly steel makers such as Nippon Steel and Kobe Steel. In 2005, around 65 per cent of the retail market had been deregulated (Miyamoto et al. 2012: 32–33). According to Scalise (2004), phase two (2000) expanded the scope of liberalization measures to include the following steps:

- Retail choice for large-lot customers: Competition was opened to utilities servicing customers with contracts for 2 MW or greater and connected to a supply system of 2000V or higher. Such customers included owners of large-scale office buildings, industrial factories, hotels, hospitals and department stores. Unlike phase one, customers could now choose their preferred wholesale supplier, whether foreign or domestic.
- Competitive bidding process for thermal capacity acquisition: The nine incumbent power companies (excluding Okinawa) were required to place competitive bids when they purchased thermal capacity. Hydroelectric and nuclear power competition were (and are) beyond the scope of the regulatory changes.
- Profit sharing: Incumbent utilities were no longer required to submit rate case applications for tariff reductions. This served as an incentive to reduce costs voluntarily because the incumbent utility could reserve part of the increased profit margin through reduced service costs.
- Abolition of Article 21 of the Antimonopoly Law: Without this provision, discriminatory rights to railroad and utility companies (natural monopolies) were no longer upheld. Both the Japan Fair Trade Commission and METI enacted guidelines that legally prohibited any exclusionary activities inhibiting new entrants.

In 2007, plans to extend deregulation to include even ordinary residential users were shelved and introducing the unbundling of the power industry disappeared from the agenda. Consequently, the market regulation has not progressed as initially anticipated and new entrants have grown to account for less than four per cent of total demand in the electricity market. Thus, while the electricity market has partially been liberalized, the electric utilities remain *de facto* monopolies (Miyamoto et al. 2012: 32–33).

Electricity Sector Organization and Supply Structure

Japan's energy policy has been the purview of METI, which, as argued in Chapter 1, has close ties to the business community. Japan's electricity policies are managed by ANRE, part of METI. Other significant operators in the electricity market are the Atomic Power Company (APC), the first Japanese company to build a nuclear reactor in 1960, which operates four nuclear power plants and sells electricity to the local power companies and the Electric Power Development Company (J-POWER), formerly a state-owned enterprise that was privatized in 2004. J-POWER operates 16 GW of hydroelectric and thermal power plants. It has also been involved in consulting services for electricity production and environmental protection in 63 countries, mainly in the developing world, since 1960 (EIA 2012b).

Among METI's chief private-sector allies are the ten regional utility monopolies: Hokkaido, Tohoku, Tokyo, Chubu, Hokuriku, Kansai, Chugoku, Shikoku, Kyushu and Okinawa. These are investor-owned, vertically integrated companies, each of which supplies customers with electric power on a retail basis in its service area (Asano 2006). The utilities organize their common interest via the Federation of Electric Power Companies of Japan (FEPC), or *Denjiren*. Since 1951, these utilities have largely monopolized control over Japan's major electricity-usage regions and, together with wholesale suppliers, currently collectively produce over 95 per cent of the country's electricity. Wholesale electric power producers with a capacity of 2 GW or above, such as the government-owned J-POWER and the APC, sell electric power to the utilities on a wholesale basis (long-term bilateral contract) (Asano 2006).

In Japan, electricity is sold to the end-consumer at a fixed price regardless of the technology used to generate electricity. The government assumes a regulatory role in ensuring that the retail price is established at a level that will not exact undue hardship on energy consumers. This system precludes Japan's utilities from strategically managing retail energy costs. With fixed retail prices, profitability for Japan's utilities comes down to cost control – minimizing the cost of power generation, maximizes gross margins. Given the importance of cost control for profitability, utilities are incentivized to favour the cheapest technologies. Nuclear power, coal-fired power and LNG-fired power represent the cheapest sources of electricity in Japan. Accordingly, it should come as no surprise that prior to the Fukushima disaster these three technologies have supplied over 80 per cent of Japan's electricity. Given their regional monopoly status, the utilities charge much higher electricity prices than those in the US and Europe (Hosoe 2006, also see Schoppa 2006: 121–23 for more detail). In fact, Japanese electricity prices have been very high by world standards and do a poor job of signalling the resource costs of meeting the demands of consumers, or the value of supply to potential producers (Hartley 2000).

Japan's electric utilities remain vertically integrated into all levels of the electricity supply chain (FEPC 2012), effectively controlling the country's regional transmission and distribution infrastructure. The largest power company is TEPCO, which accounts for 27 per cent of total power generation in the country. Since the early 2000s, the utilities have also expanded their business as new gas market entrants and, in cooperation with oil and city gas companies, are now involved in direct natural gas sales, LNG import terminals, domestic gas pipelines and international long-term supply contracts (Miyamoto 2008: 140–41). They overlap the private sector to which they belong (and, to a degree, lead) and the public sector by which they are regulated but on which they exert significant influence. The Japanese utilities have been important actors in the shaping of the country's energy policy (Gale 1981: 105). For over 60 years, Japan's utilities have enjoyed monopoly control over the power generation supply chain from resource acquisition to power generation to transmission and distribution. Under the existing system, the utility business plan is straightforward: (1) strive to negotiate the best possible retail energy prices with government regulators and (2) seek to reduce operating costs by simultaneously utilizing the least expensive energy generation technology and investing in cost minimization research.

A glance at the historical power generation by ten utilities reveals that thermal and nuclear power have grown in importance since the 1980s (Figure 8.1). Hydropower production, while significant, has remained constant in absolute levels; while geothermal power along with other renewable energy (not included in the figure) has remained at a very low level.

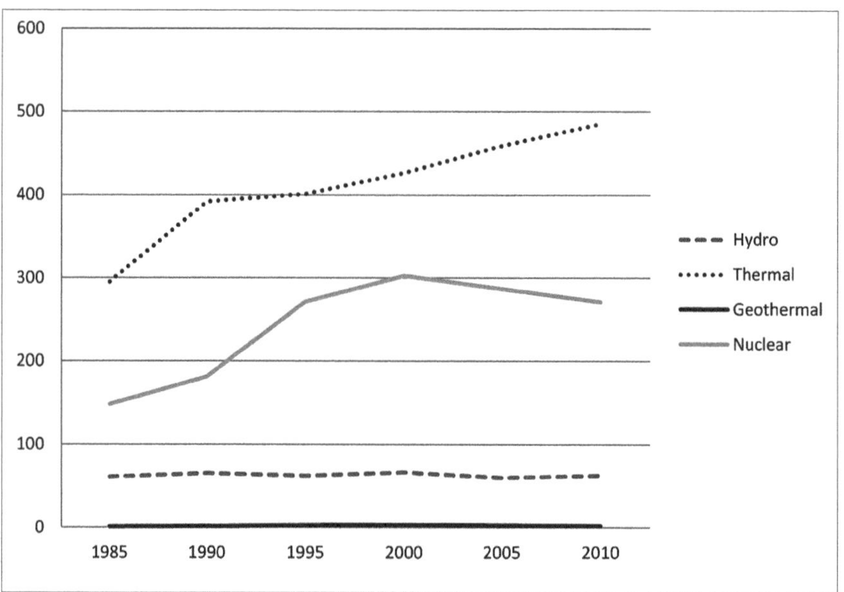

Figure 8.1 Electric power generation by ten utilities (1985–2010; TWh)
Source: FEPC 2012: 20.

Japan's electric power industry operates approximately 1,800 hydroelectric, thermal, nuclear and other power plants. In 2011, Japan had about 182 GW of installed conventional thermal electric generating capacity, including 58 thermal power plants with capacity of over 1 GW (FEPC 2012: 16). There are six additional thermal plants under construction: three using LNG and three using coal for generation. Thermal electricity generation stood at 750 TWh in 2011, a significant increase from 637 TWh in 2010 (EIA 2012b). In 2012, thermal power generation increased by additional 20 per cent, accounting for 88 per cent of Japan's electricity supply, up from 75 per cent in 2011 (IEA 2013a).

Several coal-fired plants experienced significant damage following the 2011 earthquake since they were located near Fukushima. Because of this factor, coal was not used as a substitute for nuclear power and actually experienced a negative growth in 2011 (EIA 2012b). Capacity utilization in gas-fired power facilities is close to 80 per cent, so increasing LNG use in the short-term is limited. The government has plans to construct more gas-fired power generators and, currently, there are three proposed gas-fired power plants with 3.4 GW of capacity scheduled to come online by 2016. The lead-time on greenfield plants is about 7 to 10 years mainly due to environmental permitting. However, TEPCO and TEP, utilities that suffered damage to their gas-fired plants in the earthquake zone, were temporarily exempted from these environmental requirements (EIA 2012b). Before the 2011 earthquake, Japanese utilities began removing oil-fired generation capacity due to higher operational costs. Unlike the more constricted capacity at gas-fired facilities, capacity utilization at oil-fired facilities has stood at less than 50 per cent. Therefore, power generators have had more room to increase usage of crude oil and fuel oil than natural gas in the short-term. Some utilities plan to bring back mothballed facilities to compensate for lost nuclear power (EIA 2012b).

Japan currently has 50 nuclear reactors with a total installed generating capacity of 46 GW, down from 54 reactors with 50 GW of capacity in 2010. Japan produced 271 TWh of nuclear-generated electricity in 2010 (FEPC 2012: 20). Before the Fukushima accident, Japan ranked as the third-largest nuclear power generator in the world behind the United States and France. However, the country has gradually lost all of its nuclear generation capacity as its facilities have been removed from service due to earthquake damage or for regular maintenance. Following the Fukushima disaster, Japan's nuclear electricity production dropped to 155 TWh in 2011 and further to an estimated 16 TWh in 2012 (IEA 2013a). There are currently three nuclear plants with 4.1 GW of capacity under construction (FEPC 2012: 17).

In 2011, Japan had installed hydroelectric generating capacity of 30–35 GW (FEPC 2012: 17). Similar to nuclear power, hydropower is a source for baseload generation in Japan because of the low generation costs and stable supply. Hydroelectric generation stood at 89 TWh in 2011, accounting for approximately 9 per cent of total net generation (IEA 2013a). The Japanese government has been promoting small hydropower projects to serve local communities through subsidies and by simplifying procedures. Wind, solar and geothermal power are being actively pursued in the country and installed capacity from these sources has increased significantly in recent years, particularly solar PV, as discussed in the previous chapter.

Partial Deregulation and Barriers to Entry

An assessment of Japan's electricity deregulation of the late 1990s suggests only tepid competition, average incumbent tariff declines and minimal corporate efficiencies. Government figures show that production by outside power retailers remains at 3.5 per cent, virtually unchanged from before the Fukushima accident, even though major corporations including NTT, Tokyo Gas and Mitsubishi have sought to enter the market (Iwata 2013). Structural barriers to market entry, such as extensive backup, wheeling and maintenance charges preclude cost-competitive access to many incumbent networks (transmission and distribution). In the current electricity supply system in Japan, utilities have monopoly rights to transmit and distribute electricity within their allotted franchise areas. In some cases, industrial firms with cogeneration capacity or local government authorities generate a modest amount of electricity for sale within the franchise area. These firms do not, however, directly supply consumers but must instead sell their output under contract to the monopoly utility. Their capacity is also a very small part of the total capacity in the franchise territory. Furthermore, lack of available land for further capacity building and stringent environmental regulations in generation also present strong obstacles to market entry. Following the *de jure* deregulation in the late 1990s, incumbent electric power companies have begun to lower tariff rates in anticipation of future market competition, but readjustments have been pre-emptive and voluntary, not forced through direct competition. On an index with 1990 set at 100, average electricity tariffs for Japan's three largest electric power companies (TEPCO, KEPCO and Chubu Electric Power) in 2002 fell to 93, suggesting only tepid market pressures pre-Fukushima (Miyamoto 2008).

Measures to introduce greater competition into the electricity sector have been underway in Japan for more than a decade. However, deregulation has had a limited impact on reducing electricity prices and attracting new market entrants. Consequently, the electricity market in Japan remains very heavily regulated, with pricing, entry and planning conducted in a centralized fashion. Prices remain above those paid in other advanced industrial countries and the sluggish pace of reform has deterred new entrants into the electricity sector. As the IEA (1999: 89) notes, 'Developing competition in generation is the main purpose of reform of the sector'. The major potential gains from reform result from reduced generating costs and prices that are more reflective of the marginal costs of production. Deregulation has the potential to significantly lower electricity prices in Japan

(Hosoe 2006: 232). Neither of these gains will be achieved without effective competition in generation.

The Japanese government has not focused on deregulation because natural monopoly has been considered inherent in the electricity industry (Hosoe 2006: 231). Due to the entry barrier in this sector, the industry has remained protected from market competition and prone to employ redundant employees and production facilities. However, the Fukushima disaster has reignited the debate on deregulation of Japan's electricity market. The disaster has raised doubts on the argument made by opponents of deregulation that end-to-end integration from generation to distribution ensured supply stability. There is now the possibility of a major change in direction toward development of a new electricity market structure, by shifting toward use of distributed generation sources powered by a variety of resources, development of smart grids to effectively reduce and control demand and the establishment of neutrality in the transmission sector by splitting it from supply chain in order to allow market mechanisms to function (Miyamoto et al. 2012: 33). The government has acknowledged the importance of having a neutral grid operator not tied to the interests of the utilities and in February 2013 formulated a plan to split nine power utilities into grid operators, power generators and power retailers. Under the plan, the separation of grid operation will take place as early as 2018 (Iwata 2013). The IEEJ (2013c: 3) predict that the implementation of this plan will be difficult as it could cause further financial deterioration for the utilities, which are already facing major financial difficulties.

Nuclear Power Shutdown and Financial Stress for the Utilities Since Fukushima

Given their reliance on nuclear power, all of Japan's power utilities reported losses since the Fukushima disaster due to higher fuel costs for thermal power generation (Inajima 2012). All of Japan's electric utility monopolies with the exception of Okinawa Electric Power Company own nuclear power plants and, prior to Fukushima, nuclear power has been an instrumental source in their electricity supply portfolios (see Table 8.1). All of the utilities prefer a marginal role for renewable energy (Scalise 2012b). Prior to Fukushima, reliance on nuclear power generation varied among nine utilities and ranged between 3 per cent and 44 per cent (EIA 2011a). Historically, in order to ensure a degree of coordination in commercializing nuclear technology, the government enlisted participation from Japan's three largest utilities (TEPCO, KEPCO and Chubu), which were tasked with developing a long-term commercialization strategy from as early as the 1950s (Sovacool and Valentine 2012: 117).

Table 8.1 Increased fuel costs for electric power utilities (¥ billion)

Electric power company	Nuclear as percentage of power generation pre-Fukushima (%)	Increase in fuel costs due to stoppage of nuclear plants		Net profit or loss for fiscal 2011	Net profit or loss for fiscal 2012
		Fiscal 2011	Fiscal 2012		
Hokkaido	44	50	150	–72.0	–114.6
Tohoku	26	260	250	–231.9	–174.2
Tokyo	28	880	1030	–781.6	–155.0
Chubu	13	250	220	–92.1	–65.0
Hokuriku	28	80	110	–5.2	–32.2
Kansai	44	420	700	–242.2	–582.0
Chugoku	3	0	80	2.4	–59.7
Shikoku	43	70	200	–9.3	–138.5
Kyushu	39	250	470	–166.3	–448.5
Total for nine companies	**25–30**	**2260**	**3210**	**–1598.2**	**–1769.7**

Source: The Asahi Shimbun 2012.

For the utilities, nuclear power is one of the preferred sources in the energy mix as it is relatively cheap. A glance at Table 8.2 reveals that nuclear power is the cheapest source of electricity in Japan, followed by coal and LNG. Renewable alternatives are considerably more expensive. In 2011, The IEEJ put the cost of nuclear electricity generation at ¥8.5/kWh taking into account compensation of up to ¥10 trillion for loss or damage from a nuclear accident. Later in the year a draft report for Enecan estimated nuclear generation costs for 2010 to be ¥8.9/kWh. This included capital costs (¥2.5), operation and maintenance costs (¥3.1) and fuel cycle costs (¥1.4). In addition, the estimate included ¥0.2 for additional post-Fukushima safety measures, ¥1.1 in policy expenses and ¥0.5 for dealing with future nuclear risks. The ¥0.5 for future nuclear risks is a minimum: the cost would increase by ¥0.1 for each additional ¥1 trillion of damage. The calculation for nuclear power does not include reprocessing costs for nuclear fuel or insurance liability in the event of accident damages that taxpayers would likely be obliged to bear in some form (Johnston 2011). The costs for fossil fuel generation, which factor in CO_2 abatement measures, range from ¥9.5 for coal to ¥10.7 for LNG and ¥22.1 for oil. Forward projections to 2030 indicate that the nuclear cost remains stable but fossil fuel costs increase significantly (World Nuclear Association 2012).

Table 8.2 Power generation cost in Japan for major energy sources (¥ per kWh)

<table>
<tr><th></th><th>Nuclear</th><th>Coal</th><th>LNG</th><th>Oil</th><th>Solar</th><th>Wind</th><th>Geothermal</th><th>Hydro</th></tr>
<tr><td>METI</td><td>5.0–6.0</td><td>...</td><td>7.0–8.0</td><td>...</td><td>49.0</td><td>10.0–14.0</td><td>8.0–22.0</td><td>...</td></tr>
<tr><td>IAE</td><td>4.8–6.2</td><td>5.0–7.1</td><td>5.7–7.1</td><td>10.0–17.3</td><td>37.0–46.0</td><td>10.0–14.0</td><td>...</td><td>8.2–13.3</td></tr>
<tr><td>TEPCO</td><td>6.1</td><td colspan="3">9.1</td><td colspan="3">30.5</td><td>...</td></tr>
<tr><td>Scalise</td><td>...</td><td>8.8</td><td>13.1</td><td>19.6</td><td>30.0–45.0</td><td>14.0</td><td>10.0</td><td>...</td></tr>
<tr><td>ACNRE</td><td>5.3</td><td>5.7</td><td>6.2</td><td>10.7</td><td>...</td><td>...</td><td>...</td><td>11.9</td></tr>
<tr><td>Matsuo et al.</td><td>7.2</td><td colspan="3">10.2</td><td>...</td><td>...</td><td>8.9</td><td>...</td></tr>
<tr><td>FEPC</td><td>5.1–7.4</td><td>6.0–7.6</td><td>8.4–10.1</td><td>9.0–15.0</td><td>30.0–58.7</td><td>10.0–15.0</td><td>8.0–22.0</td><td>8.0–13.0</td></tr>
<tr><td>IEEJ</td><td>8.9</td><td>9.5</td><td>10.7</td><td>22.1</td><td>33.4–38.3</td><td>9.4–23.1</td><td>9.2–11.6</td><td>...</td></tr>
</table>

Source: Dale 2011, IAE 2009, IEEJ 2012a (for Cost Verification Committee of the Central Economic Council), METI 2011a, Scalise 2011a, Scalise 2012a, Matsuo et al. 2011 (for the Subcommittee to Study Costs and Other Issues in the government's Advisory Committee for Natural Resources and Energy).

Given the relatively low cost of nuclear power, rising energy costs due to the nuclear shutdown and increased costs of fossil fuel imports after Fukushima have had a negative effect on utilities' profitability. As a result, eight electric utilities have incurred losses for fiscal year 2011 (Hosoe 2012b: 53) and nine utilities are forecast to have incurred financial loses for 2012 (see Table 8.1). Moreover, with nuclear reactors idle, they have become stranded assets that require costly maintenance, adding additional financial pressure on the utilities (Interview with Scott Valentine, 17 January 2013). In 2012, the utilities posted a ¥150 billion cost for depreciation, which relates to recovery of nuclear-plant construction expenses and for the maintenance of idled reactors.

The utilities have been spending heavily on increased procurements of LNG and crude oil to sustain higher output from thermal power plants. Fuel costs for the 10 companies amounted to ¥3.4 trillion in the first six months of 2012, up 40 per cent on an annual basis. In late 2012, Hokkaido, Tohoku, Shikoku, Kansai and Kyushu suggested that they would raise electricity prices by 14–19 per cent for corporate customers, following the example of TEPCO, which increased prices by 15 per cent in September 2012. While the government has raised some concerns about increased power rates, the move seems inevitable given the deregulation of electricity prices. While the government expects utilities to conduct the utmost cost cutting before they can propose rate hikes, it has no power to intervene in pricing issues between corporate customers and utilities because of the liberalization of these markets. Historically, since their profit margin has been fixed, the utilities have had no incentive to purchase fossil fuels at lower prices compared with other countries. They knew that they could automatically pass their increased fuel costs on to consumers. Moreover, price deregulation has left the utilities in full control of the power grid, ultimately blocking attempts by outside companies to gain a foothold in the market. Most of Japan's corporate customers have had no choice but to accept the proposed rate increases because of the virtual monopoly enjoyed by regional utilities despite a nominal liberalization of the sector in the mid-1990s (*The Asahi Shimbun* 2012, Iwata 2013).

In their survey of energy security perceptions in Japan, Valentine et al. (2011) found that most Japanese citizens are acutely aware of the need to minimize energy costs both to support industrial competitiveness and household energy expenditure. Their preferences for affordable energy services temper any policies that might commit Japan to costly low-carbon technological transition initiatives in the energy sector (Valentine et al. 2011). Any increased cost of electricity caused by the uptake of renewable energy will be distributed among consumers, who already pay among the highest electricity prices in the world. This is a clear indication that there may be little public support or economic incentive for utilities to move away from traditional energy sources to renewable energy.

Against this backdrop, the electric utilities have been pushing for urgent nuclear power restarts in order to improve their financial standing. In fact, a respondent suggested that if nuclear reactors are not restarted, 'several utilities will go bankrupt' (Interview with Akira Miyamoto, 10 January 2013). As discussed in Chapter 1 and Chapter 6, these deep-pocketed monopolies have cultivated salubrious ties with influential politicians through generous campaign contributions (Duffield and Woodall 2011). Nine of Japan's electric utilities spent a combined ¥2.4 trillion to sponsor TV programs and run ads in print media over four decades to promote nuclear power and underscore the safety of their plants (*The Asahi Shimbun* 2013). Their size, *de facto* monopoly position, control over pricing data and privately owned assets put them at an advantage to comparable companies in most other industrial democracies (Scalise 2012b). Lobbyists from large power utilities have in the past opposed more ambitious renewable energy goals. They have substantial influence at the local and national governmental levels (Ferguson 2011). Given the relative cost of nuclear power, any future plan to downsize or eliminate nuclear energy is certain to face considerable opposition from the utilities.

Deficiencies in the Electricity Grid

A specific challenge for Japan, that became apparent during the energy crisis after Fukushima, is the need for power grid alignment between electric power companies and between eastern and western Japan. The country is divided into ten regional monopolies with weak inter-grid connections and Japan does not run its grid on the same nationwide frequency, the western part, with the total capacity of 154 GW, running on 60Hz and the eastern, with the capacity of 122 GW, on 50Hz. The frequency difference is a result of a historical 'accident' from the nineteenth century, when Tokyo's electrical entrepreneurs installed 50Hz generators mainly from Germany, while their counterparts in Osaka selected 60Hz equipment from the United States. The result is a national grid whose two halves can only exchange 1 GW of power. In terms of weak inter-grid connections between franchise areas, TEPCO can receive at most 7.2 per cent of its peak demand from neighbouring utilities. KEPCO and Chubu can receive at most 19.5 per cent and 9.5 per cent of their peak demand from their neighbours, respectively. The isolated utilities, such as Kyushu and Hokkaido, with links only to Chugoku and TEP, can only receive at most 1.0 per cent and 6.7 per cent of their peak demand, respectively. The utility most exposed to potential competition is Chugoku, where almost 33.5 per cent of the peak demand could potentially be supplied by the neighbouring utilities.

While the power grid mis-alignment and the lack of inter-regional connections is not an issue under normal circumstances, it became a major problem in the aftermath of the Fukushima disaster. As an immediate result of the Fukushima disaster, about 9.7 GW of nuclear power capacity went offline in the region, along with about 9 GW of additional thermal power capacity. The utilities affected were TEPCO and TEP. TEP was unable to provide the Kanto region with additional power, because its power plants were damaged in the earthquake. KEPCO was unable to share electricity, because its system operates at 60Hz, whereas TEPCO and TEP operate their systems at 50Hz. Two substations, one in Shizuoka Prefecture and one in Nagano Prefecture, were able to convert between frequencies and transfer electricity from Kansai to Kanto and Tōhoku, but their capacity was limited to around 1 GW (Hosoe 2011). This frequency difference between eastern and western Japan has limited TEPCO's ability post-Fukushima to seek help from the 56 per cent of Japan's power-generating capacity that lies to the west.

The calls for stronger links between various regions and eastern and western Japan date back more than a decade (Hartley 2000: 22). According to Tanaka (2012), western Japan has the greater number of power generators, so efforts should be made, by steadily upgrading the oldest equipment first, to standardize power supply at western Japan's 60Hz. However, the issue for Japan is that the basic structure of the transmission system cannot be changed easily since it is the result of long-run optimization reflecting regional characteristics (Asano 2006). Creating new linkages between the 50Hz and 60Hz systems is expensive and would require a long time to materialize. However, in January 2013, the nine electric utilities announced that they would augment the facilities, or frequency converter stations, for grid interconnection between eastern and western Japan from the current capacity of 1 GW to 2.1 GW. The augmentation project is estimated to cost ¥132 billion to ¥141 billion, which the nine utilities will recover in the form of additional wheeling fees. With planned start of operation in 2020, the utilities will start work on the upgrade during 2013. They will seek cooperation from the national government in order to gain assistance in legal procedures and negotiations with landowners for facility construction (*The Denki Shimbun* 2013).

Another electricity grid challenge for Japan is the lack of historical commitment and funding directed towards the establishment of a smart electricity grid. A smart grid works by combining advanced communication, sensing and metering infrastructure with the existing electricity network. Smart grids have enormous potential to improve the efficiency of Japan's electricity sector and transform the way Japanese use energy in households and commercial sectors. A smart grid can improve the reliability of electricity services for consumers by identifying and resolving faults on the electricity grid, better managing voltage and identifying infrastructure that requires maintenance. Smart grids can also help consumers manage their individual electricity consumption and enable the use of energy efficient 'smart appliances' that can be programmed to run on off-peak power. Smart grid technology helps to convert the power grid from static infrastructure that is operated as designed to flexible and environmentally friendly infrastructure that is operated proactively. In Japan, 'smart grid' implies energy transmission and distribution to promote the stability of the electric power supply, by using information and communication technology while introducing a high level of renewable energy (Poh et al. 2012).

Historically, METI's perspective on the smart grid has been shaped through regulatory capture by the utilities (DeWit 2011). In February 2009, the Vice Minister of METI stated that Japan does not need a smart grid. According to DeWit (2011), Japan's monopolized utilities have been satisfied with the status quo of their highly balkanized and inflexible power network. They have been unwilling to innovate, unless those innovations maintained the status quo that has been lucrative for the nuclear village. They understood that a truly smart grid would be a serious threat to their monopolies by, among other things, encouraging deregulation and the entry of serious competition. In the past, in circumstances when Japan sought to develop a smart grid, the innovative potential was blunted by potent vested interests seeking to control their scope (DeWit 2011).

Unsurprisingly then, prior to the Fukushima disaster, smart grids were not seen as priorities for the Japanese political economy. The nuclear village wanted to sell nuclear technology, domestically and overseas and received commitments from the national government. The electric monopolies were also not interested in setting up smart grids because this would have involved lower power demand (and fewer nuclear reactors) and would have also helped the renewable producers that had long been trying to gain access to the grid (DeWit 2011). However, following the Fukushima disaster, Japan's electricity market has been undergoing a smart revolution. Following Fukushima, both METI and NEDO have been supporting numerous smart grid research and development projects (Poh et al. 2012).

In 2009, Yokohama, Japan's second-largest city launched a small-scale environmental experiment, encouraging residents to install solar panels on their roofs and purchase costly equipment to track how much energy they use. Yokohama officials' goal was simple: to save power and cut the city's carbon emissions. But since the nuclear disaster that transformed the way Japan thinks about both energy and the companies that supply it, Yokohama's 'smart city project' has taken on potentially larger significance. What began as a modest environmental plan now stands as a controversial blueprint for a system in which the country's monopolistic utilities would lose their absolute control of the grid (Interview with Andrew DeWit, 16 January 2013). In Yokohama, the households with both solar panels and meters act as micro-size power companies, generating electricity, using the required amount and in some cases selling the surplus back to TEPCO. That model contrasts sharply with the one that has served Japan for decades, as ten privately owned utility companies established regional fiefdoms, largely reliant on coastal nuclear plants and allowing little room for renewable energy projects that would cut into profits. Although nuclear and renewable energy are not mutually exclusive, many investors and industry analysts blame the frailty of Japan's renewable energy sector on the long dominance of the utility companies and their cohorts, including the heavy machinery giants that build nuclear plants and a pro-nuclear bureaucracy in Tokyo (Harlan 2011).

The smart revolution centres on mixing information technology with the electricity transmission grid. Through applying IT to the grid, as in the case of Yokohama, consumers are becoming able to adjust their own power consumption and power production through applications on their mobile phones as well as other devices. Electric companies are learning how to accurately monitor power consumption and production in real time. Among other benefits, this will allow them to handle the flux of power from renewable sources, achieve greater efficiency and reduce the massive amounts of generating capacity needed to cope with peak demand. This interactive smart grid project has already seen the diffusion of smart meters (DeWit 2011), with Japan expected to install nearly 80 million smart meters in the next decade (Berst 2012). The future challenge for Japan will be on how to stabilize power supplies nationwide as large amounts of wind and solar power start entering the grid and to increase awareness about the benefits of the smart grid among the Japanese people (Poh et al. 2012).

Conclusion

This chapter demonstrates that despite *de jure* electricity market liberalization in the 1990s, Japan's electricity market has remained *de facto* controlled by ten regional monopolies. The main issue is that the utilities remain in control of both electricity generation and transmission, which presents an enormous obstacle to potential new entrants. However, following the Fukushima disaster, the government has acknowledged the importance of having a neutral grid operator not tied to the interests of the utilities. In February 2013, the government formulated a plan to split nine power utilities into grid operators, power generators and power retailers, under which the separation of grid operation will take place by 2018. At the same time, the chapter demonstrates that, following Fukushima, electric utilities have been affected by the nuclear power shutdown and have found themselves in a difficult financial position as they are bearing the increased cost of energy imports to fuel thermal power plants. The utilities, as part of the iron triangle, remain committed to nuclear power and have historically remained opposed to an increased penetration of renewable energy due to high costs. This speaks to the protracted nature of energy transitions as one of the sources of path dependency in Japan's energy mix. With the utilities opposed to a greater penetration of renewable energy due to its relatively high costs, in the absence of government intervention, it will be extremely challenging for Japan to transition away from nuclear power and fossil fuels. The chapter also highlights the fragmented nature of Japan's electricity grid, which was a major issue in the aftermath of the Fukushima disaster and future plans to improve the grid.

Conclusion
Japan's Future Energy Options

Japan is currently struggling to implement energy policy to counter recurring shocks reinforced by the Fukushima disaster. Against this backdrop, the conclusion evaluates the feasibility of various energy policy options for Japan against the backdrop of the Fukushima disaster. However, it is important to begin by restating the major underlying premise of this book: Japan's future energy policy and energy mix are path dependent. Policy responds less directly to social and economic conditions than it does to the consequences of past policy (Hall 1993: 277). While one may argue that Japan needs to start from scratch after Fukushima in redrafting its energy policy, this is not feasible, given the embedded interests and the prevailing market arrangements that will require a generational shift to change. In this context, the book examined the evolution of Japan's historical energy policy and energy policy-making structures in order to place current challenges in the appropriate context. As evident from the analysis of each energy source in the preceding chapters, Japan now faces multiple and often multidimensional challenges. While most of Japan's energy challenges are not new, many of these challenges have been exacerbated by the Fukushima disaster and, consequently, have made them more urgent. In other words, while many of these challenges existed prior to the Fukushima disaster, the disaster magnified their intensity and immediacy with which they need to be tackled.

Will the Fukushima disaster and the ensuing energy crisis lead to structural change in Japan's energy policy and institutions that govern energy? Various Japan specialists suggested pre-Fukushima that changes of a more systemic and transformative nature in Japan's political economy seem likely (Schoppa 2006, Vogel 2006, Schaede 2008). However, Witt (2006) argues that the coordinated nature of Japan's business system hinders adjustment to changing economic activity. Witt (2006) examines the symptoms of business sclerosis in Japan and queries why the forces of inertia are so strong. Witt surmises that the reason for this was that institutional rigidity in the Japanese business system rendered it relatively slow to react to changes impinging on economic activity. In particular, he identifies three sources of such rigidity: the widespread support for the extant system; the significance of existing institutions as a source of competitive advantage; and the coordinated nature of the business system, facilitated by extensive social networks. With the exception of an increase in foreign ownership and the concomitant unwinding of shareholdings by Japanese financial institutions, there is scant evidence of divergence from pre-bubble conditions in Japan. Since Japanese firms and individuals tend to be risk-averse, this has adverse implications for institutional change (Witt 2006).

Witt (2006) suggested that Japan is unlikely to change substantially until it is faced with a viable alternative, or, more forcefully, until it has no choice at all. Japan would have to encounter a true crisis before it would reform. Arguably, the Fukushima disaster is one such crisis and a critical juncture, which represents both a challenge and an opportunity for Japan. Crises are generative, as they release long-held preferences from policy entrepreneurs and open them to public scrutiny and debate. In the past, Japan has experienced most change on those rare occasions when it has been subjected exogenously, such as during the oil crises of the 1970s. What distinguished Japan from other industrial nations in its response to the 1973 oil crisis were the intensity, breadth and determination of its strategic response to the transformation in international energy and economic realities. Yet while a crisis can catalyse public opinion and crystallize debate, it does not force change on its own. In Fukushima, Japan has suffered its biggest external energy shock since the 1973 energy crisis. This external shock has provided a window of opportunity for change. However, given developments to date explored throughout this book, it is unlikely that Fukushima will result in structural policy and institutional change.

More specifically, in the aftermath of the Fukushima disaster, one idea that has gained traction is that the ensuing energy crisis is likely to induce a corresponding change in the institutions that govern Japan's energy industry. The challenge for Japan is to remove the insider–outsider dichotomy in terms of access to energy policy-makers. According to Kingston (2012a: 204), the great risk Japan faces is not failing to recover from Fukushima, or taking on too much debt, but rather pursuing a recovery that looks too much like the past. If the nation only recovers to regain the level of stagnation and inertia that prevailed pre-Fukushima, it will be an indictment of the ruling elite and a negative barometer for Japan's future. According to DeWit et al. (2012: 156), the most salient political economy question after Fukushima is the degree to which vested interests can minimize reform to the energy economy and its policy-making institutions. It has been suggested that the system of energy policy-making in Japan needs a reform away from a mode that caters to vested interests and toward a more transparent and socially fair policy. The energy crises cannot be allowed to reinforce and recycle the same forces and structures that enabled the crisis (Shadrina 2012: 77).

However, the vested interests centred on the nuclear village remain in control of key positions of power and a move away from nuclear power seems unlikely, public opinion notwithstanding. Although the public expressed overwhelming support for the nuclear phase-out once Enecan released the three options for Japan's future energy policy in July 2012, post-Fukushima developments indicate that those who hold power oppose this option. Although increasingly vocal, for example through *The Japan Times*, critics remain marginalized as their voice is hardly heard in the mainstream media and they lack agency to affect structural change both in terms of energy policy and institutional overhaul. Regardless of societal pressures to move away from nuclear power, Japan's future energy policy remains embedded in the country's institutional and organizational structure, with the METI as the energy policy-making hub, the nuclear industry, the utility monopolies and the LDP party apparatus at the centre. This energy policy-making structure has remained remarkably stable for almost four decades and remains in place despite of the societal pressure to move away from nuclear power since the Fukushima disaster.

This post-Fukushima reality does not look particularly bright for meaningful nuclear policy reform. It is exceedingly difficult to break the intimate relationship between cohesive state nuclear regulatory agencies and to succeed in preventing the vested interests from regaining control over the nuclear regulating agencies and creating institutional rigidities favouring the status quo (Shadrina 2012: 81). The bureaucracy remains beholden to the nuclear village, *amakudari* and *amaagari* continue to see officials transition from positions of power within the bureaucracy and there is a dearth of policy entrepreneurs who seek to challenge the status quo. The vested interests remain entrenched, perhaps even more so with the return to power of the LDP. Since Fukushima, there is a consensus among energy policy-makers that a greater level of oversight is required to ensure that the country's nuclear power operators are held to higher standards of accountability. At the same time, however, the influential nuclear village remains embedded in the policy-making process (Sovacool and Valentine 2012).

The most immediate challenge for Japan and the one which will affect how the country tackles many other challenges, is to arrive at a decision regarding the future of nuclear power. Given the power of the nuclear village and the Abe government's recognition of the necessity of nuclear power to the Japanese economy, it is possible that some of Japan's idle nuclear reactors will be restarted in 2014. An early indication of this was a July 2013 statement from Kenzo Oshima, NRA commissioner, who told *Reuters* that 'some units are projected (to restart) one year from now, though I don't know how many' (*Reuters* 2013). In November 2013, the NRA announced that it has no fixed schedule to complete safety checks at idled nuclear power plants, possibly delaying reactor restarts. As of late 2013, there are 14 plants which the electric power companies consider in compliance with the new regulatory requirements, that took effect on 8 July and for which they have applied for restart (Adelman and Suga 2013). The echelons of power in Tokyo believe that restarting the reactors deemed safe by the NRA under new and improved safety standards is the only choice for Japan. Their cause is strengthened by Japan's precarious energy security situation in the aftermath of the disaster, which is once again used as justification for continued reliance on nuclear power. Under these circumstances, the vested interests still look potent enough to deflect meaningful nuclear reforms.

Consequently, one could argue that, in times of upheaval and uncertainty, policy-makers and businessmen are just as likely to steer a course that reinforces the status quo and place their trust in institutions that have endured and served Japan well in the past energy crises. This does not necessarily imply that they are against change. However, such change may be incremental, arising from the organic, slow evolution of society (Mahoney and Thelen 2010). Policy-makers are likely to be in a stronger position to resist pressure from societal interests when they are armed with a coherent policy paradigm (Hall 1993: 190). In the case of Japan's energy policy, this paradigm centres on nuclear power as an integral part of Japan's future energy mix. While this paradigm does not dictate the optimal course for policy for all stakeholders, at least it provides a set of criteria for resisting some societal demands while accepting others.

The energy angst (or anxiety) of Japan's elite institutions, both public and private, which is rooted in Japan's radical lack of energy resources, has generated a distinct, embedded heritage that is likely to continue shaping Japan's energy policies in the future as it has in the past, shifting political party configurations notwithstanding (Calder 2012: 183–84). Despite the rigidities and resource misallocations, that embedded angst has generated dramatic improvements in Japanese energy security since the oil shocks of the 1970s, as Japan has sharply diversified away from oil and greatly reduced the energy intensiveness of the national economy. Given these historical contingencies, the appeal of nuclear energy for Japan, the continued dominance of vested interests and the increasing costs of imported fossil fuels, the energy crisis is unlikely to lead to the phase-out of nuclear power, at least not in the immediate term. In particular, retaining nuclear power in Japan's energy mix would allow for a well-diversified energy portfolio (Interview with Hayden Lesbirel, 28 January 2013).

Historically, Japan's nuclear policies and even Japan's energy policies have been characterized by the 'carry-out-government-plans' formula and the dual system of 'decided by the state and operated by business'. Fukushima has made the ease with which the utilities get away with regulatory fraud and the collusion between the utilities and the regulator, abundantly clear. If Japan is to continue relying on nuclear power – and this appears very likely – for whatever share of electricity production, such practice has to change and it will be essential to regain people's confidence, mainly through independent functioning of the newly established nuclear regulator. The fair assessment of the future of nuclear power can be ensured only through a democratic and transparent decision-making process. In theory, energy security should be considered on a national level where democratic practice implies respect and wishes of the public. At the same time, the government should publicly express economic considerations that may prevent or inhibit the development of a specific energy source. However, the nature and structure of Japan's politics and political economy both pre- and post-Fukushima, in which bureaucrats, the LDP and business interests remain in control of public discourse and policy and regulatory processes, suggests that Japan is not a true liberal democracy.

Consequently, it is difficult to be optimistic about the prospects for bold action on urgently needed reforms beyond establishment of a new nuclear regulating agency. The evidence presented in this book suggests that reform – exemplified by the symbolic restructuring of nuclear regulators and establishment of the NRA – has been piecemeal and that the embedded interests are likely to prevail in the long-term. Nuclear policy change and a more transparent and democratic policy process will be difficult to achieve while the vested interests remain in control.

Although there has been considerable change in the shape of the Japanese nuclear policy-making arena over the past fifty years, such change overwhelmingly has been in the direction of the further strengthening and augmentation of the number of veto players, such as METI and the electric utilities, making it ever harder to bring about a radical break from the state's traditional nuclear policies (Hymans 2011). Historically, the existence of numerous veto players implies that Japan could not make a radical nuclear policy shift. Yet, civil society and local government opposition to siting of nuclear power plants have in some cases prevented the government from moving ahead with its ambitious targets in the past. In the wake of Fukushima, this opposition has become increasingly robust, with civil society not only opposed to siting, but also to the continued presence of nuclear power in some regions of the country. Moreover, the emergence of a new veto player—the NRA—has implied that the nuclear village no longer has carte blanche to act with unfettered power. In line with Mahoney and Thelen (2010), the emergence of the NRA, in tandem with increased local government opposition, has led to a weakening in the ability of the nuclear village to defend the status quo. Nevertheless, we have not witnessed broad-scale structural change in Japan's nuclear policy or political economy that would suggest that Fukushima will be a critical juncture that will lead to the wholesale abandonment of nuclear power and the breakdown of the old vestiges of power. The findings of this book reinforce Mahoney and Thelen's (2010) thesis that institutional change occurs gradually and through the cumulation of seemingly small adjustments.

What *may* cause a radical change in Japan's embedded power structures and nuclear policy? While such a scenario remains unlikely for the foreseeable future, three developments could result in increased impetus for change. First, Japan may move away from nuclear power if the country is affected by another nuclear disaster, similar in scale to Fukushima, or if a similar disaster occurs internationally. Such a disaster may result in mass public protests, much larger in scale than that seen after Fukushima, which could provide sufficient pressure for policy change. Second, a demographic and generational shift, together with continued economic malaise over the next decade, may result in increased pressure for change in the structure of Japan's politics and economy, where a new generation of politicians, business leaders and other public figures with alternative views of Japan's future, may hold sufficient agency to implement significant change. Indeed, as Samuels (2013) suggests, such agency for change has been gathering momentum at the local government level since Fukushima. Finally, if renewable energy gains sufficient share of the electricity market and if it achieves economies of scale that make it cost-competitive with nuclear power, Japan's energy policy-makers and electric utilities may recognize that the energy security rationale for continued reliance on nuclear power are no longer prevalent.

The discussion about the future of nuclear energy and regulatory reform in Japan is unfolding across and overlapping with an immense number of related issues, such as the economics of renewable energy and climate change policy, electricity market reform and energy efficiency. Moe (2007) argues that in order to enable the destruction of old inefficient industries and promote potentially promising industries – such as renewable energy industry – the role of the state is to ensure that no vested interests become so powerful and influential that they can effectively hold veto power over any substantial reform. The Japanese state has not played this role and it is unlikely that this status quo will change. At the same time, the government and the utilities have not perceived renewable energy as a potentially promising industry, which, as a consequence, has remained on the margins of energy policy-making.

It is important to recognize that energy policy demands political-strategic thinking that takes into account political variables as well as economic ones. In fact, METI has a long history of prioritizing energy security above prices (Schoppa 2006: 127). Japan's experience in alternative energy policy after 1973 suggests that there are circumstances in which government intervention is warranted because market operations can lead to sub-optimal policy outcomes, such as in the case of overreliance on oil prior to the oil crises. Significant government intervention in the energy markets is necessary in order to support greater penetration of renewable energy and, consequently, to accelerate Japan's energy transition. Without government intervention, Japan's energy mix is likely to remain on its business-as-usual trajectory. In this context, the findings of this book highlight the protracted nature of energy transitions as a source of path dependency in Japan's energy mix, a condition in which locked-in and cheaper or, more importantly those energy sources that have significant political support for whatever reason, continue to dominate. The renewable energy industry has been characterized by high levels of risk and uncertainty associated with long developmental lead times and high capital costs. Consequently, electric utilities have been unwilling to bear the full costs involved in developing renewable energy capacity even though it may be in the longer-term national interest to do so. Therefore, it is not surprising that Japan has lagged behind its competitors in terms of renewable energy penetration.

Under these conditions, government policies – such as the new FIT – seeking to facilitate longer-term renewable energy development and minimize over-reactions to short-term fluctuations in energy prices may be more optimal from a broader policy perspective than policies, which allow market forces to operate in an unconstrained way. In fact, Japan's future energy strategy could reduce supply vulnerability if it strengthens capacity for energy technology innovation. A combination of favourable economics and supportive government policies can have an incubator-like effect, stimulating investment in renewable technology. The incubator effect is most likely to emerge if there is a stable policy environment, as this would provide incentives for long-term planning and investment. These conditions encourage private firms to build economies of scale and make incremental improvements to renewable technologies, which in turn reduce costs and improve reliability. Energy security benefits can be realized and energy transition accelerated if new technology draws on domestically available energy, reducing the need for imported fossil fuels.

Yet, while large scale penetrations of photovoltaic, wind and geothermal power can replace a part of nuclear power in future power generation system in Japan, it will be very difficult to remove nuclear power entirely considering the impacts from the economic, environmental and energy security perspective. It is impossible to force a more costly fuel into the system to displace a cheaper one without reducing economic activity potential. The long-term commitment to renewable energy is likely to result in severe consequences for the already struggling economy, with higher electricity prices making Japanese corporations less competitive and fuelling the movement of jobs offshore. Moreover, even with renewable energy reaching 20 per cent of overall energy supply within a decade – a highly ambitious target – a significant supply gap will remain if no new nuclear power plants are to be built and the existing plants remain idle. Consequently, in such a scenario, Japan's reliance on imported fossil fuels is likely to increase, resulting in higher costs and emissions and further exacerbating traditional security of supply issues in the context of competitive environment in Northeast Asia. A nuclear phase-out in Japan would result in an 8 per cent increase in CO_2 emissions by 2040 (Nakata 2002). The main impact of a nuclear phase-out would be an increase in fossil fuel consumption in the electricity sector. As nuclear power is phased out, the output of coal and gas-fired power generation would grow proportionally, leading to an increase in emissions. It is apparent that achieving significant reductions in GHG emissions without nuclear power will be more costly than with nuclear power.

In terms of electricity market reform, the Japanese government should reassess the value of preserving Japan's regional electricity monopolies, who some have blamed for relatively high electricity prices. One of the key issues that the government will need to consider when formulating its energy policy is the already high cost of energy in Japan and the adverse impact this has had on the competitiveness of the Japanese economy and which has already contributed to the trend for Japan's heavy industry to relocate overseas. Japan's regional power companies dominate the electricity business from generation to transmission and distribution, with a *de facto* monopoly over supply to households in their areas of operation and a near-monopoly to commercial customers. The government should engage in a comprehensive reassessment of this structure and the future operations of the power industry while developing its new energy policy. The government should consider removing the existing power-supply boundaries in the short to medium-term, which would effectively end regional monopolies by allowing newcomers to enter the power business so that consumers have the option to choose among power suppliers.

Valentine (2011: 6843–44) argues that Japan's energy policy network lacks the sufficient diversity to avoid misguided strategic decisions. More specifically, the vertical integration of Japan's electric utilities into all levels of the electricity supply chain has been a key hurdle to the diffusion of renewable energy. Accordingly, the electricity generation industry should be fully liberalized. Currently a form of 'Japanese' liberalization exists whereby utilities have monopoly control over all but small specialized segments of the national energy supply chain. Operation of the electricity grid should be clearly separated from the energy generation function and power generation should be open to competition in order to ensure that generation costs are minimized.

It is commendable that Japan is in fact moving in this direction, with reforms announced in mid-2012. In July 2012, the Expert Committee on Electric Power System Reforms approved the measures outlined in the interim report entitled *Basic Policy on the Electric Power System – For a Power System Open to the Public*. The Committee agreed that an environment that facilitates real competition would be created by fully opening retail market and eventually abolishing tariff restraints under which electric company's profit is added to the cost. Moreover, restrictions on wholesaling will be abolished in line with the full opening of the retail market. To ensure supply stability, retail suppliers will be required to secure a certain supply capacity and a capacity market will be established to trade the surpluses and deficits in supply capacity. A mechanism for procuring power supply and recovering costs will be established to address the long-term supply–demand gap. In electricity transmission and distribution, a 'wide-area grid operator' will be set up to coordinate the transmission and distribution beyond the borders of supply areas (IEEJ 2012e: 6). It is now crucial that Japan embarks on these reforms.

With regard to economic efficiency, according to Moe (2012), most of the energy efficiency improvements in Japan occurred during the first two decades of this drive. Duffield and Woodall (2011) argue that the easiest gains to energy efficiency have already been made and there are particular challenges ahead for Japan to develop technologies, which may further improve energy efficiency. While they acknowledge that there may be significant efficiency gains to be made, the industrial sector may not be the best place to look for them. This indicates that Japan has made so much progress in deploying existing energy efficiency technologies that there are no energy-conserving opportunities left for its production process. To be sure, industry remains the largest energy consumer, at 46 per cent in 2008 (EDMC 2010: 38). However it has also been the principal target of government efforts to increase energy efficiency since the 1970s – approximately 90 per cent of the energy consumption in the sector has long been covered by the *Energy Conservation Law* and, partly as a result, the share of energy consumption attributable to the industrial sector has steadily declined, from nearly two-thirds in 1973 (EDMC 2010).

Despite remarkable improvements to energy efficiency in the industrial sector, Moe (2012) argues that further improvement in energy efficiency is possible and that the progress has been slow over the past two decades due to the power of vested interests. While making further strides in energy efficiency has been the major strategy of most LDP administrations (DeWit and Tani 2008), the iron triangle of LDP, METI and the business associations (primarily *Keidanren*) has allowed for very lax emissions regulations. For energy security reasons METI has been a strong proponent of energy efficiency, but with two decades of economic stagnation, a harmony of interests with industrial interest groups has led to a hollowing-out of energy efficiency policies. Consequently, energy efficiency improvements have slowed to a crawl, as the economy ground into depression and cutbacks were made to clean energy investments. While Japan is still one of the leaders, for two decades there has been little progress, allowing other countries to catch up. The economic output relative to energy consumption in various European countries is now higher than in Japan. Currently, Japan's super-efficient high-tech export industries live side-by-side with an inefficient service sector. Looking forward, in the household, office and transportation sectors, the absolute consumption amounts can be reduced by introducing generous subsidies by 2020 for the diffusion of energy-saving equipment and facilities (household appliances and automobiles, housing, etc.) (Sawa 2012: 135, Interview with Shinichi Kihara, 17 January 2013, Interview with Hideaki Fujii, 11 January 2013). Energy conservation for buildings has much room for improvement compared to Europe or the US (IEEJ 2013a: 3) and introduction of smart grids, as discussed in Chapter 8, may be a promising way forward.

Another major challenge for Japan is that following the Fukushima disaster, the country has increased its fossil fuel imports in order to replace lost nuclear power. Japan's aggressive search for more fossil fuels has exacerbated some of the old challenges related to fossil fuels and is diametrically opposed to Japan's energy policies since the 1970s oil crises. Increased demand for and cost of imported fossil fuels has not only had a negative impact of Japan's trade balance, but is also taking place in the context of increased competition for energy in the Asia–Pacific region.

While it is apparent that energy security in Japan is no longer considered only in terms of oil supply, Japan remains over-reliant on the Middle East and, while its large stockpiles provide a safety net during a potential supply crisis, this overreliance leaves Japan vulnerable to price volatility and panic buying if the Iranian situation escalates into an outright conflict. A rapid increase in oil prices would not only make oil imports more expensive, but given that LNG prices are linked to oil prices in Asia, Japan's LNG import bill would increase significantly from its current record level. Hence, Japan's strategy to challenge the existing LNG pricing structures in the region is of utmost priority for the government and this was confirmed by a respondent from METI (Interview with Shinichi Kihara, 17 January 2013). It is likely that, with this goal in mind, Japan and other Asian LNG importers will import significant volumes of North American LNG within a decade. It remains to be seen whether North American LNG will be a game changer in the international LNG market. A recent strategy that Japan has employed (since early 2013) in order to reduce demand for and associated costs of imported LNG has been to increase thermal coal imports. While the international coal market is well supplied, Japan remains over-reliant on Australia – particularly for thermal coal – and this overreliance leaves it vulnerable to potential supply disruptions in Australia due to floods or other natural disasters. It also leaves Japan vulnerable to the direction of annual bilateral negotiations with Australian coal exporters. Moreover, it is paramount that Japan continues to develop CCS technology and further improve efficiency of its coal-fired power plants in order to reduce GHG emissions from burning increasing volumes of coal.

Japan's increased demand for imported fossil fuels could not have occurred at a less convenient time. The major dilemma before Japan over the past decade has been how to deal with its energy-hungry Asian competitors with whom the possibility of cooperation is limited (Jain 2007: 39). When the NNES was introduced in 2006, the threat from China has been described as a key driver (Koike 2006: 45). The NNES and Japan's other energy policy documents have limited the scope for cooperation with energy importing nations in Asia. Whether deliberate or not, omitting mention of cooperation with regional rivals to secure energy supply from Russia and other regions including the Middle East is clearly indicative of Japan's stance on this issue (Choo 2009: 52).

Looking ahead, despite a recent surge in demand, it is clear that Japan will constitute an ever smaller share of Asia's energy supply and demand mix. To recognize the adjustment in thinking driven by these changes, it is important to understand Japan's historical energy position. For decades, Japan dominated Asia's energy picture. It had the region's largest manufacturing base, largest number of vehicles and an active consumer culture, all of which drove it to be the region's largest energy consumer and fuel importer. This picture no longer holds true, particularly for oil. By 2012, Japan's oil demand was 4.7 million bpd – representing just over 15 per cent of regional demand. It has been overtaken by China, which has emerged as an increasingly important centre of demand over the past decade. In 2002, China's oil demand reached parity with Japan. By 2012, China's oil demand of 10.2 million bpd was more than twice that of Japan (BP 2013). In addition, as its incremental take has tapered off and as other countries have entered the market for gas, Japan's relative share in the world's LNG market has begun to decline. Moreover, China recently overtook Japan as the world's largest coal importer. Over the next 15 years these trends are expected to accelerate.

Japan's move toward more government intervention in energy markets since 2006 carries risks. Powerful domestic interests have used energy security as a rationale to back policies and projects that consume large amounts of resources, but have contributed relatively little to the country's energy security, evidenced in continued inability to meet self-developed oil targets. Since Japan's actions are not isolated, a move away from the market has induced others in the region to act similarly. Energy security is now a vital national security concern for major powers in Asia and energy nationalism dominates their behaviour (Lam 2009). Moreover, a more interventionist Japan complicates – and perhaps even jeopardizes – efforts by the United States to ensure that all countries in the region abide by a common set of rules and norms for energy-related trade and investment.

As a result, the current trajectory of energy markets in the Asia–Pacific is consistent with Robert Jervis' security dilemma. Jervis (1978) argued that security for one state reduces the security of the other. According to this 'security dilemma', many of the means by which a state tries to increase its security, decrease the security of others. However, an energy security gain for one state does not necessarily need to be a loss for other states. The expanded conceptualization of energy security employed in this book illustrates the entire set of interests at stake and thereby identifies the areas where interests overlap. Consistent with Keohane and Nye (2001), the potential exists under these conditions for states to build institutions in order to lower transaction costs and pursue absolute rather than relative gains. Recognizing this 'energy security dilemma' is a first step towards mitigation of energy security cooperation problems in the Asia–Pacific region (Vivoda 2010: 5262).

In Northeast Asia, no institutionalized multilateral security mechanism exists. China and Japan are crucial actors for multilateral energy security cooperation to be successful. Jaewoo Choo (2009) argues that Northeast Asian governments remain highly distrustful of one another and this has limited the prospects for energy security cooperation in the region. He notes that while China and Japan do engage in general energy cooperation in technical fields, both governments have little interest in energy security cooperation, since such energy issues are linked to the governments' narrow perspectives on the concept of national security. Beijing and Tokyo prefer to pursue their energy security strategies separately. They have competed with each other over access to international oil and gas in zero-sum terms. China and Japan employ bilateral deals to enhance their energy security and this tendency is increasingly reflective of their desires to 'lock in' access to scarce resources in what is increasingly perceived to be a competitive environment (Wilson 2012: 434–40). The issue of regional energy security cooperation in the context of Northeast Asian regional formation is symbolic of the Sino-Japanese competition for regional leadership in Northeast Asia.

Kent Calder (1997) warned that energy competition in Asia will lead to strategic rivalry and represents a recipe for conflict. Consequently, China and Japan need to recognize the fact that their lack of energy cooperation due to mutual political distrust will not only impair their own energy security, but may also have negative implications on regional stability. Consequently, the two powers need to create a more positive political atmosphere and promote bilateral political trust. It could be a hard task for China and Japan to do so under their current confrontation, but their cooperation, rather than rivalry, will be essential not only to the development of the bilateral relationship, but also to regional stability and energy cooperation in Northeast Asia. In particular, Japan should demonstrate awareness of the implications of its actions for stable energy supply for the region as a whole and exercise leadership in building a framework for regional cooperation in energy.

Scope for cooperation between Asian nations exists, due to their joint interest in securing SLOCs from war and piracy, thus enabling an uninterrupted flow of energy. Political stability of key regions where energy sources originate and safe passage of SLOCs are in the interest of major powers in Asia and elsewhere. Manning's (2000) potential areas for cooperation include cross-border natural gas pipelines; electricity grid link-ups; joint activities in fighting maritime piracy and in establishing sea-lane security; cooperation in nuclear energy and the management of nuclear waste. Japan should help China to pursue a more effective policy in energy efficiency and conservation. In this context, Japan's experience in restricting oil use, support for alternative fuels and energy efficiency is relevant to China (Calder 2012: 184–85). With some of the most efficient and environmentally friendly technologies in the world, particularly related to lower GHG emissions and increased efficiency of coal-fired power plants, Japan could share its technology with other countries. Likewise, China is one of the world's leaders in wind technology and Japan may have a lot to gain from cooperation with China in this field. Both states' energy interests could be achieved at lower cost if they could cooperate. Certainly, there is little benefit to Japan of continuing to compete against China for access to global oil and gas reserves. Cooperating with China for access to global resources, in exchange for assistance with energy efficiency technologies, should have resonance in Tokyo. This could lead to a way out of the zero-sum perspective in a region where there is little tangible cooperation on energy security (Liao 2009, Vivoda 2010). Cooperation in energy will help to foster mutual confidence and trust and the hope that similar sentiments might spread within the political sphere is not a vision that should be dismissed lightly as sheer optimism. Regional energy security in the Asia–Pacific requires a multilateral approach. It is in Asia–Pacific nations' interests that they pool their resources together and jointly strive for collective energy security. For that reason, multilateral initiatives are preferable to unilateral or bilateral efforts and China and Japan are key regional states to encourage multilateral cooperation.

Until the Fukushima disaster, every policy statement of the Japanese government since the 1970s has stressed Japan's dependence on imported energy sources and the need to manage the risks associated with that structural dependence to ensure Japanese security. The Fukushima disaster highlights the importance of planning for possible supply disruptions of domestic origin. The key is to ensure that Japan's domestic energy infrastructure is upgraded in order to be able to cope with any similar future disruptions. Particular attention needs to be given to electricity grid upgrades and extension of the gas pipeline network in order to enhance the resilience of Japan's domestic energy system to future shocks. An electricity grid upgrade and, in particular, equalizing the electricity grid to either 50Hz or 60Hz would also support a more rapid penetration of renewable energy. According to Hiroshi Hamasaki from Fujitsu Research Institute, equalizing the voltage in the grid and upgrading the existing network would allow for a six-fold increase in wind power supply over the next decade (Interview with Hiroshi Hamasaki, 16 January 2013).

Energy security, the environment and the economy have long been the three pillars of Japanese energy policy. Japan's future energy policy cannot overlook any of these three pillars. Yet, as discussed throughout the book, there are significant challenges associated with each of the pillars and the government is in an extremely difficult position of finding the best policy with which to tackle a multitude of interconnected challenges over the next decade. In fact, the book demonstrates that no energy source in Japan, with a notable pre-Fukushima exception of nuclear power, is perfect in terms of 3Es (energy security, economic cost and environmental effect). While abundant, fossil fuels are becoming increasingly costly and emissions intensive; while environmentally acceptable, renewable energy is prohibitively expensive; and finally, while nuclear power has been considered affordable, reliable and low emissions source of energy prior to Fukushima, the disaster illustrates that nuclear power can be expensive and extremely harmful to the environment if not managed effectively. Consequently, it is likely that Japan's future energy policy will be shaped around an additional pillar – safety (Interview with Shinichi Kihara, 17 January 2013). If safety of Japan's nuclear power stations can be assured, only then can it regain some of the lost appeal among Japan's public.

Although the results of Japan's energy policy transition have yet to create widespread popular conviction that the new energy policy promotes the hallowed 3E formula, the changes that have occurred have been piecemeal, but are nevertheless noticeable. These include the establishment of an independent nuclear regulator (with the jury still out on the degree of its independence), the passing of the new and improved FIT and planned reforms in the electricity sector. It is important now that the Japanese government maintains the transition momentum so all feasible reforms can reinforce each other and create a genuinely comprehensive, coherent and efficient new energy policy. In the long run, the government should strive for an integrative policy framework, balancing priorities of energy security, environmental policy, economic policy as well as safety. The government's response will set the course with regards to enhancing energy security, reducing energy costs and carbon emissions and enhancing nuclear reactor safety. Whether and how these objectives can be reconciled will depend on whether political consensus can be reached.

List of Respondents

Prof Ken Koyama, Managing Director and Chief Economist, The Institute of Energy Economics, Japan

Mr Akira Ishii, Senior Visiting Researcher (Oil and Gas Business Environment), Oil and Gas Upstream Business Unit, Japan Oil, Gas and Metals National Corporation

Dr Akira Miyamoto, General Manager (Executive Researcher), Strategy Planning and Research Team, Energy Resources and International Business Unit, Osaka Gas Co. Ltd.

Mr Toyoshi Matsumoto, Manager, Research and Planning Team, Planning Department, Energy Resources and International Business Unit, Osaka Gas Co. Ltd.

Ms Chikako Ishiguro, Senior Analyst, Research and Planning Team, Planning Department, Energy Resources and International Business Unit, Osaka Gas Co. Ltd.

Dr Hiroshi Hamasaki, Research Fellow, Economic Research Centre, Fujitsu Research Institute

Mr Shinichi Kihara, Director, Internal Affairs Division, Agency for Natural Resources and Energy

Dr Paul J. Scalise, JSPS Research Fellow, Institute of Social Science, The University of Tokyo

A/P Scott Valentine, Associate Director, MPP/IP Program, Graduate School of Public Policy, The University of Tokyo

A/P Ben McLellan, Graduate School of Energy Science, Kyoto University

A/P Hideaki Fujii, Faculty of Economics, Kyoto University

Prof Andrew DeWit, College of Economics, Department of Economic Policy Studies, Rikkyo University

A/P Hayden S. Lesbirel, Coordinator of Political Science in the School of Arts and Social Sciences, Faculty of Arts, Education and Social Sciences, James Cook University.

Distinguished Professor Emeritus Vaclav Smil, Faculty of Environment, University of Manitoba.

Bibliography

Books

Ackerman, T. 2005. *Wind Power in Power Systems*. Hoboken, NJ: John Wiley & Sons.

Aldrich, D.P. 2008. *Site Fights: Divisive Facilities and Civil Society in Japan and the West*. Ithaca, NY: Cornell University Press.

Ayres, C.E. 1944. *The Theory of Economic Progress*. Chapel Hill: University of North Carolina Press.

Beder, Sharon. 2003. *Power Play: The Fight to Control the World's Electricity*. New York: W.W. Norton & Company.

Berger, T.U. 1998. *Cultures of Antimilitarism: National Security in Germany and Japan*. Baltimore, MD: Johns Hopkins University Press.

Blyth, M. 2002. *Great Transformations: Economic Ideas and Institutional Change in the Twentieth Century*. Cambridge: Cambridge University Press.

Boyle, G. 2004. *Renewable Energy: A Power for a Sustainable Future*. Oxford: Oxford University Press.

Bradford, T. 2006. *Solar Revolution*. Cambridge: MIT Press.

Calder, K.E. 1997. *Asia's Deadly Triangle: How Arms, Energy and Growth Threaten to Destabilize Asia–Pacific*. London: Nicholas Brealey Publishing.

Calder, K.E. 2012. *The New Continentalism: Energy and Twenty-First-Century Eurasian Geopolitics*. New Haven, CT: Yale University Press.

Castells, M. 2000. *End of Millennium*. Oxford, UK: Blackwell.

Cole, B.D. 2008. *Energy Security in Asia: Sea Lanes and Pipelines*. Westport, CT: Praeger Security International.

Colignon, R.A. and Usui, C. 2003. *Amakudari: The Hidden Fabric of Japan's Economy*. Ithaca, NY: Cornell University Press.

Culter, S. 1999. *Managing Decline: Japan's Coal Industry Restructuring and Community Response*. Honolulu, HI: University of Hawaii Press.

Dargin, J. and Lim T.W. 2012. *Energy, Trade and Finance in Asia*. London: Pickering & Chatto.

Dorian, J.P. 1995. *Energy in China: Foreign Investment Opportunities, Trends and Legislation*. London: Financial Times Energy Publishing.

Harris, C. 2006. *Electricity Markets: Pricing, Structures and Economics*. Hoboken, NJ: John Wiley & Sons.

Hein, L.E. 1990. *Fueling Growth: The Energy Revolution and Economic Policy in Post-war Japan*. Cambridge, MA: Harvard University Press.

Horsnell, P. 1997. *Oil in Asia: Markets, Trading, Refining and Deregulation.* London: Oxford University Press.

IEA (International Energy Agency). 1999. *Energy Policies of IEA Countries – Japan: 1999 Review*. Paris: OECD/IEA.

IEA (International Energy Agency). 2003. *Energy Policies of IEA Countries – Japan: 2003 Review*. Paris: OECD/IEA.

IEA (International Energy Agency). 2008. *Energy Policies of IEA Countries – Japan: 2008 Review*. Paris: OECD/IEA.

IEA (International Energy Agency). 2011a. *Key World Energy Statistics*. Paris: OECD/EIA.

Johnson, C. 1982. *MITI and the Japanese Miracle*. Stanford, CA: Stanford University Press.

Katz, R. 2003. *Japanese Phoenix*. New York: M.E. Sharpe.

Keohane, R.O. and Nye, J.S. 2001. *Power and Interdependence.* Third Edition. New York: Longman.

Lesbirel, S.H. 1998. *NIMBY Politics in Japan: Energy Siting and the Management of Environmental Conflict*. Ithaca, NY: Cornell University Press.

Low, M., Nakayama, S. and Yoshioka, H. 1999. *Science, Technology and Society in Contemporary Japan*. Cambridge: Cambridge University Press.

Mahoney, J. and Thelen, K. 2010. *Explaining Institutional Change: Ambiguity, Agency and Power*. Cambridge University Press, New York, NY.

Manning, R.A. 2000. *The Asian Energy Factor: Myths and Dilemmas of Energy, Security and the Pacific Future.* New York: Palgrave.

March, J. and Olsen, J. 1989. *Rediscovering Institutions*. New York: Free Press.

McKean, M. 1981. *Environmental Protest and Citizen Politics in Japan*. Berkeley, CA: University of California Press.

Moe, E. 2007. *Governance, Growth and Global Leadership: The Rise of the State in Technological Process*. Aldershot and Burlington, UK: Ashgate.

Mokyr, J. 1990. *The Lever of Riches*. Oxford: Oxford University Press.

North, D. 1990. *Institutions, Institutional Change and Economic Performance*. Cambridge: Cambridge University Press.

Olson, M. 1982. *The Rise and Decline of Nations*. London: Yale University Press.

Sakaiya, T. 1975. *Yudan!* ("*Oil Cut-off?*") Tokyo: Nihon Keizai Shinbunsha (*in Japanese*).

Sakakibara, E. 2003. *Structural Reform in Japan.* Washington, DC: Brooking Institution Press.

Samuels, R.J. 1987. *The Business of the Japanese State: Energy Markets in Comparative and Historical Perspective*. Ithaca, NY: Cornell University Press.

Samuels, R.J. 1994. *Rich Nation, Strong Army*. Ithaca, NY: Cornell University Press.

Samuels, R.J. 2013. *3.11: Disaster and Change in Japan*. Ithaca, NY: Cornell University Press.

Schaede, U. 2008. *Choose and Focus: Japanese Business Strategies for the 21st Century*. Ithaca, NY: Cornell University Press.

Schoppa, L.J. 2006. *Race for the Exits: The Unravelling of Japan's System of Social Protection*. Ithaca, NY: Cornell University Press.

Schreurs, M. 2002. *Environmental Policy in Japan, Germany, and the United States*. Cambridge: Cambridge University Press.

Schumpeter, J.A. 1942. *Capitalism, Socialism and Democracy*. New York: Harper Torchbooks.

Schumpeter, J.A. 1983 (1934). *The Theory of Economic Development*. London: Transaction Publishers.

Smil, V. 2008. *Energy in Nature and Society*. Cambridge: MIT Press.

Smil, V. 2010. *Energy Transitions: History, Requirements, Prospects*. Santa Barbara, CA: Praeger.

Sovacool, B.K. and Valentine, S.V. 2012. *The National Politics of Nuclear Power: Economics, Security, and Governance*. New York: Routledge.

UNDP (United Nations Development Programme). 2004. *World Energy Assessment, 2004 Update*. New York: UNDP.

Vernon, R. 1983. *Two Hungry Giants: The United States and Japan in the Quest for Oil and Ores*. Cambridge, MA: Harvard University Press.

Vogel, S.K. 2006. *Japan Remodelled: How Government and Industry Are Reforming Japanese Capitalism*. Ithaca, NY: Cornell University Press.

Wizelius, T. 2007. *Developing Wind Power Projects: Theory and Practice*. London: Earthscan.

Wu, Y.L. 1977. *Japan's Search for Oil: A Case Study on Economic Nationalism and International Security*. Stanford, CA: Hoover Institution Press.

Book Chapters

Aldrich, D.P. 2012b. Networks of power: institutions and local residents in post-Tōhoku Japan, in *Natural Disaster and Nuclear Crisis in Japan: Response and Recovery after Japan's 3/11*, edited J. Kingston. New York: Routledge, 127–39.

Basrur, R., Koh, S.L.C. and Chang, Y. 2012. Post-Fukushima: whither nuclear energy?, in *Nuclear Power and Energy Security in Asia*, edited by R. Basrur and S.L.C. Koh. New York: Routledge, 194–202.

Broadbent, J. 2002. Japan's environmental regime, in *Environmental Policy in Industrialized Countries*, edited by U. Desai. Cambridge, MA: MIT Press, 295–355.

Caldwell, M. 1981. The dilemmas of Japan's oil diplomacy, in *The Politics of Japan's Energy Strategy*, edited by R.A. Morse. Berkeley, CA: University of California Press, 65–84.

Choo, J. 2009. Northeast Asia energy cooperation and the role of China and Japan, in *Energy and Security Cooperation in Asia: Challenges and Prospects*, edited by C. Len and E. Chew. Stockholm: Institute for Security and Development Policy, 41–60.

Cohen, L., McCubbin, M. and Rosenbluth, F.M. 1995. The politics of nuclear power in Japan and the United States, in *Structure and Policy in Japan and the United States*, edited by P. Cowhey and M. McCubbins. Cambridge, MA: Cambridge University Press, 177–202.

DeWit, A. and Kaneko, M. 2011. Moving out of the "nuclear village", in *Tsunami: Japan's Post-Fukushima Future*, edited by J. Kingston. Washington, DC: Foreign Policy, 213–24.

DeWit, A. and Tani, T. 2008. The local dimension of energy and environmental policy in Japan, in *Japanstudien 20. Regionalentwicklung und Regionale Disparitäten*, edited by V. Elis and R. Lützeler. München: Iudicium Verlag, 281–305.

DeWit, A., Iida, T. and Kaneko, M. 2012. Fukushima and the political economy of power policy in Japan, in *Natural Disaster and Nuclear Crisis in Japan: Response and Recovery after Japan's 3/11*, edited by J. Kingston. New York: Routledge, 156–71.

Donnelly, M.W. 1993. Japan's nuclear energy quest, in *Japan's Foreign Policy after the Cold War: Coping with Change*, edited by G.E. Curtis. London: M.E. Sharpe, 179–201.

Egenhofer, C. 2013. The growing importance of carbon pricing in energy markets, in *The Handbook of Global Energy Policy*, edited by A. Goldthau. Oxford: John Wiley & Sons, 358–72.

Flower, A. 2008. Appendix 2, LNG pricing in Asia – Japan Crude Cocktail (JCC) and S-curves, in *Natural Gas in Asia: The Challenges of Growth in China, India, Japan, and Korea*, edited by J. Stern. New York: Oxford University Press, 405–9.

Fujii, H. 2000. Japan, in *Rethinking Energy Security in East Asia*, edited by P.B. Stares. Tokyo: Japan Center for International Exchange, 59–78.

Gale, R.W. 1981. Tokyo Electric Power Company: its role in shaping Japan's coal and LNG policy, in *The Politics of Japan's Energy Strategy*, edited by R.A. Morse. Berkeley, CA: University of California Press, 85–105.

Hall, P.A. and Soskice, D. 2001. An introduction to varieties of capitalism, in *Varieties of Capitalism: The Institutional Foundations of Comparative Advantage*, edited by P.A. Hall and D. Soskice. Oxford: Oxford University Press, 1–70.

Hashimoto, K., Elass, J. and Eller, S.L. 2006. Liquefied natural gas from Qatar: the Qatargas project, in *Natural Gas and Geopolitics: From 1970 to 2040*, edited by D.G. Victor, A.M. Jaffe and M.H. Hayes. Cambridge: Cambridge University Press, 234–67.

Hayes, M.H. and Victor, D.G. 2006. Politics, markets, and the shift to gas: insights from the seven historical case studies, in *Natural Gas and Geopolitics: From 1970 to 2040*, edited by D.G. Victor, A.M. Jaffe and M.H. Hayes. Cambridge: Cambridge University Press, 319–56.

Iida, T. 2011. Country perspective: Japan, in *The End of Nuclear Energy?*, edited by N. Netzer and J. Steinhilber. Berlin: Friedrich-Ebert-Stiftung, 48–52.

Ikuta, T. 1980. The energy issue and the role of nuclear power in Japan, in *Australia and Japan: Nuclear Energy Issues in the Pacific*, edited by S. Harris and K. Oshima. Canberra and Tokyo: Australia–Japan Economic Relations Research Project, 19–28.

Itoh, S. 2009. Russia's energy policy towards Asia: opportunities and uncertainties, in *Energy and Security Cooperation in Asia: Challenges and Prospects*, edited by C. Len and E. Chew. Stockholm: Institute for Security and Development Policy, 143–66.

Itoh, S. 2012. Sino-Japanese competition over Russian oil, in *Eurasia's Ascent in Energy and Geopolitics: Rivalry or Partnership for China, Russia and Central Asia?*, edited by R.E. Bedeski and N. Swanström. New York: Routledge, 158–78.

Jaffe, A.M., Hayes, M.H. and Victor, D.G. 2006. Conclusions, in *Natural Gas and Geopolitics: From 1970 to 2040*, edited by D.G. Victor, A.M. Jaffe and M.H. Hayes. Cambridge: Cambridge University Press, 467–83.

Jain, P. 2007. Japan's energy security policy in an era of emerging competition in the Asia–Pacific, in *Energy Security in Asia*, edited by M. Wesley. New York: Routledge, 28–41.

Keohane, R.O. 1993. Institutional theory and the realist challenge after the Cold War, in *Neorealism and Neoliberalism, The Contemporary Debate*, edited by D.A. Baldwin. New York: Columbia University Press, 269–300.

Kim, J.D. and Byrne, J. 1996. The Asian atom: hard-path nuclearization in East Asia, in *Governing the Atom: The Politics of Risk*, edited by J. Byrne and S.M. Hoffman. London: Transaction Publishers, 271–97.

Kingston, J. 2012a. The politics of disaster, nuclear crisis and recovery, in *Natural Disaster and Nuclear Crisis in Japan: Response and Recovery after Japan's 3/11*, edited by J. Kingston. New York: Routledge, 188–206.

Lam, P.E., 2009. Japan's energy diplomacy and maritime security in East Asia, in *Asian Energy Security: The Maritime Dimension*, edited by H. Lai. London: Palgrave Macmillan, 115–34.

Leaver, R. 2009. Australia's role in feeding Asia's energy demand, in *Energy and Security Cooperation in Asia: Challenges and Prospects*, edited by C. Len and E. Chew. Stockholm: Institute for Security and Development Policy, 121–42.

Liao, X. 2009. Perceptions and strategies on energy security: the case of China and Japan, in *Energy and Security Cooperation in Asia: Challenges and Prospects*, edited by C. Len and E. Chew. Stockholm: Institute for Security and Development Policy, 105–20.

Miyamoto, A. 2008. Natural gas in Japan, in *Natural Gas in Asia: The Challenges of Growth in China, India, Japan, and Korea*, edited by J. Stern. New York: Oxford University Press, 116–73.

Moriguchi, C. 1988. Japan's energy policy during the 1970s, in *Government Policy towards Industry in the United States and Japan*, edited by J.B. Shoven. Cambridge: Cambridge University Press, 301–18.

Morita, Y. 2010. Energy saving in the Japanese industrial sector and international cooperation, in *Energy Security: Asia Pacific Perspectives*, edited by V. Gupta and G.K. Ching. New Delhi: Manas Publications, 85–108.

Morse, R.A. 1981a. Energy and Japan's national security, in *The Politics of Japan's Energy Strategy*, edited by R.A. Morse. Berkeley, CA: University of California Press, 37–64.

Morse, R.A. 1981b. Japan's energy policies and options, in *The Politics of Japan's Energy Strategy*, edited by R.A. Morse. Berkeley, CA: University of California Press, 1–14.

Morse, R.A. 1982. Japanese energy policy, in *After the Second Oil Crisis Energy Policies in Europe, America and Japan*, edited by W.L. Kohl. Lexington, MA: Lexington Books, 255–70.

Nakatani, K. 2004. Energy security and Japan: the role of international law, domestic law, and diplomacy, in *Energy Security: Managing Risk in a Dynamic Legal and Regulatory Environment*, edited by B. Barton, C. Redgwell, A. Rønne and D.N. Zillman. Oxford: Oxford University Press, 413–27.

Nye, J.S. 1981. Japan, in *Energy and Security*, edited by D.A. Deese and J.S. Nye. Cambridge, MA: Ballinger Publishing Company, 211–28.

Orren, K. and Skowronek, S. 1994. Beyond the iconography of order: notes for a new institutionalism, in *The Dynamics of American Politics*, edited by L.C. Dodd and C. Jillson. Boulder, CO: Westview, 311–32.

Oyama, K. 1998. The policymaking process behind petroleum industry regulatory reform, in *Is Japan Really Changing Its Ways? Regulatory Reform and the Japanese Economy*, edited by L.E. Carlile and M.C. Tilton. Washington, DC: The Brookings Institution, 142–62.

Petri, P.A. 1981. High-cost energy and Japan's international economic strategy, in *The Politics of Japan's Energy Strategy*, edited by R.A. Morse. Berkeley, CA: University of California Press, 15–36.

Sawa, A. 2012. Conflicting policies: energy security and climate change policies in Japan, in *Energy Security in the Era of Climate Change*, edited by L. Anceschi and J. Symons. New York: Palgrave Macmillan, 126–42.

Scalise, P.J. 2004. National energy policy: Japan, in *Encyclopedia of Energy*, Volume 4, edited by C.J. Cleveland. San Diego, CA: Elsevier Academic Press, 159–71.

Scalise, P.J. 2011b. Can TEPCO survive?, in *Tsunami: Japan's Post-Fukushima Future*, edited by J. Kingston. Tokyo: Foreign Policy, 204–12.

Scalise, P.J. 2012a. Hard choices: Japan's post-Fukushima energy policy in the twenty-first century, in *Natural Disaster and Nuclear Crisis in Japan: Response and Recovery after Japan's 3/11*, edited by J. Kingston. New York: Routledge, 140–55.

Stewart, D. 2009. Japan: the power of efficiency, in *Energy Security Challenges for the 21st Century: A Reference Handbook*, edited by G. Luft and A. Korin. Santa Barbara, CA: Praeger Security International, 176–90.

Sudo, S. 2008. Energy security challenges to Asian countries from Japan's viewpoint, in *Energy Security: Visions from Asia and Europe*, edited by A. Marquina. New York: Palgrave Macmillan, 147–61.

Suttmeier, R.P. 1981. The Japanese nuclear power option: technological promise and social limitations, in *The Politics of Japan's Energy Strategy*, edited by R.A. Morse. Berkeley, CA: University of California Press, 106–33.

Thomson, E. 2009. Southeast Asia's energy and security challenge, in *Energy and Security Cooperation in Asia: Challenges and Prospects*, edited by C. Len and E. Chew. Stockholm: Institute for Security and Development Policy, 21–40.

Yokobori, K. 2005. Japan, in *Energy & Security: Toward a New Foreign Policy Strategy*, edited by J.H. Kalicki and D.L. Goldwyn. Washington, D.C.: Woodrow Wilson Center Press, 305–28.

Journal Articles

Aldrich, D.P. 2011. Future fission: why Japan won't abandon nuclear power. *Global Asia*, 6(2), 62–67.

Asano, H. 2006. Regulatory reform of the electricity industry in Japan: What is the next step of deregulation? *Energy Policy*, 34(16), 2491–97.

Atsumi, M. 2007. Japanese energy security revisited. *Asia–Pacific Review*, 14(1), 28–43.

Ayoub, N. and Yuji, N. 2012. Governmental intervention approaches to promote renewable energies – special emphasis on Japanese feed-in tariff. *Energy Policy*, 43, 191–201.

Birol, F. 2013. Coal's role in the global energy mix: treading water or full steam ahead? *Cornerstone: The Official Journal of the World Coal Industry*, 1(1), 6–9.

Bó, E.D. 2006. Regulatory capture: a review. *Oxford Review of Economic Policy*, 22(2), 203–25.

Bobrow, D.B. and Kudrle, R.T. 1987. How middle powers can manage resource weakness: Japan and energy. *World Politics*, 39(4), 536–65.

Calder, K.E. 2013. Beyond Fukushima: Japan's emerging energy and environmental challenges. *Orbis*, 57(3), 438–52.

Chester, L. 2010. Conceptualizing energy security and making explicit its polysemic nature. *Energy Policy*, 38(2), 887–95.

Chow, L.C. 1992. The changing role of oil in Chinese exports, 1974–89. *The China Quarterly*, 131, 750–65.

Dauvergne, P. 1993. Nuclear power development in Japan: "outside forces" and the politics of reciprocal consent. *Asian Survey*, 33(6), 576–91.

DeCarolis, J.F. and Keith, D.W. 2006. The economics of large-scale wind power in a carbon constrained world. *Energy Policy*, 34(4), 395–410.

Drifte, R. 2005. Japan's energy policy in Asia: cooperation, competition, territorial disputes. *Oil, Gas & Energy Law*, 3(4), no page numbers.

Duffield, J.S. and Woodall, B. 2011. Japan's new basic energy plan. *Energy Policy*, 39(6), 3741–49.

Dusinberre, M. and Aldrich, D.P. 2011. Hatoko comes home: civil society and nuclear power in Japan. *The Journal of Asian Studies*, 70(3), 683–705.

Eguchi, Y. 1980. Japanese energy policy. *International Affairs*, 56(2), 263–79.

Fukai, S. 1988. Japan's energy policy. *Current History*, 87(528), 169–84.

Funabashi, Y. and Kitazawa, K. 2012. Fukushima in review: a complex disaster, a disastrous response. *Bulletin of the Atomic Scientists*, 68(2), 9–21.

Gasparatos, A. and Gadda, T. 2009. Environmental support, energy security and economic growth in Japan. *Energy Policy*, 37(10), 4038–48.

Gilpin, R. 1996. Economic evolution of national systems. *International Studies Quarterly*, 40(3), 411–31.

Goldstein, L. and Kozyrev, V. 2006. China, Japan and the scramble for Siberia. *Survival*, 48(1), 163–78.

Golob, S. 2003. Beyond the policy frontier: Canada, Mexico, and the ideological origins of NAFTA, *World Politics*, 55(3), 361–98.

Gorges, M.J. 2001. The new institutionalism and the study of the European Union: the case of the social dialogue. *West European Politics*, 24(4), 152–68.

Greener, I. 2001. Social learning and macroeconomic policy in Britain. *Journal of Public Policy*, 21(2), 133–52.

Grübler, A. 1991. Energy in the 21st century: From resource to environmental and lifestyle constraints. *Entropie*, 164/165, 29–33.

Haddad, M.A. 2010. A new state-in-society approach to democratization with examples from Japan. *Democratization*, 17(10), 997–1023.

Hall, P.A. 1993. Policy paradigms, social learning, and the state. The case of economic policymaking in Britain. *Comparative Politics*, 25(3), 275–96.

Hay, C. 1999. Crisis and the structural transformation of the state: interrogating the process of change. *The British Journal of Politics and International Relations*, 1(3), 317–44.

Hayashi, M. and Hughes, L. 2013. The policy responses to the Fukushima nuclear accident and their effect on Japanese energy security. *Energy Policy*, 59, 86–101.

Hogan, J. and Doyle, D. 2007. The importance of ideas: an a priori critical juncture framework. *Canadian Journal of Political Science*, 40(4), 883–910.

Hogan, J. and Feeney, S. 2012. Crisis and policy change: the role of the political entrepreneur. *Risks, Hazards & Crisis in Public Policy*, 3(2), article 6.

Honda, T. 1998. NEDO's solar energy program. *Renewable Energy*, 15(1–4), 114–18.

Horiuchi, A. and Schimizu, K. 2001. Did amakudari undermine the effectiveness of regulator monitoring in Japan? *Journal of Banking & Finance*, 25(3), 573–96.

Hosoe, N. 2006. The deregulation of Japan's electricity industry. *Japan and the World Economy*, 18(2), 230–46.

Hosoe, T. 2005. Japan's energy policy and energy security. *Middle East Economic Survey*, 48(3), 17 January.

Huenteler, J., Schmidt, T.S. and Kanie, N. 2012. Japan's post-Fukushima challenge – implications from the German experience on renewable energy policy. *Energy Policy*, 45, 6–11.

Ikenberry, G.J. 1986. The irony of state strength: comparative responses to the oil shocks in the 1970s. *International Organization*, 40(1), 105–37.

Inoue, Y. and Miyazaki, K. 2008. Technological innovation and diffusion of wind power in Japan. *Technological Forecasting and Social Change*, 75(8), 1303–23.

Jervis, R. 1978. Cooperation under the security dilemma. *World Politics*, 30(2), 167–74.

Katzenstein, P.J. and Okawara, N. 1993. Japan's national security: structures, norms, and policies. *International Security*, 17(4), 84–118.

Kiani, B. 1991. LNG trade in the Asia–Pacific. *Energy Policy*, 19(1), 63–75.

Klein, D.W. 1980. Japan 1979: the second oil crisis. *Asian Survey*, 20(1), 42–52.

Koike, M. 2006. Japan looks for oil in the wrong places. *Far Eastern Economic Review*, 169(8), 44–47.

Koike, M., Mogi, G. and Albedaiwi, W.H. 2008. Overseas oil-development policy of resource-poor countries: A case study from Japan. *Energy Policy*, 36(5), 1764–75.

Kondo, K. 2009. Energy and exergy utilization efficiencies in the Japanese residential/commercial sector. *Energy Policy*, 37(9), 3475–83.

Krasner, S.D. 1984. Approaches to the state: alternative conceptions and historical dynamics. *Comparative Politics*, 16(2), 223–46.

Kumar, S., Kwon, H.-T., Choi K.-H., Cho, J.H., Lim, W. and Moon, I., 2011. Current status and future projections of LNG demand and supplies: a global perspective. *Energy Policy*, 39(7), 4097–4104.

Langton, T. 1994. LNG prospects in the Asia–Pacific region. *Resources Policy*, 20(4), 257–64.

Lesbirel, S.H. 1988. The political economy of substitution policy: Japan's response to lower oil prices. *Pacific Affairs*, 61(2), 285–302.

Lesbirel, S.H. 1990. Implementing nuclear energy policy in Japan. *Energy Policy*, 18(3), 267–82.

Lesbirel, S.H. 2004. Diversification and energy security risks: the Japanese case. *Japanese Journal of Political Science*, 5(1), 1–22.

Lesbirel, S.H. 2003. Markets, transaction costs and institutions: compensating for nuclear risk in Japan. *Australian Journal of Political Science*, 38(1), 5–23.

Lesbirel, S.H. 1991. Structural adjustment in Japan: terminating "old king coal". *Asian Survey*, 31(11), 1079–94.

Levy, J.S. 1994. Learning and foreign policy: sweeping a conceptual minefield. *International Organization*, 48(3), 279–312.

Liao, X. 2007. The petroleum factor in Sino-Japanese relations: beyond energy cooperation. *International Relations of the Asia–Pacific*, 7(1), 23–46.

Lidsky, L.M. and Miller, M.M. 2002. Nuclear power and energy security: a revised strategy for Japan. *Science and Global Security*, 10(2), 127–50.

Manicom, J. 2008. Sino-Japanese cooperation in the East China Sea: limitations and prospects. *Contemporary Southeast Asia*, 30(3), 455–78.

Maruyama, Y., Nishikido, M. and Iida, T. 2007. The rise of community wind power in Japan: enhanced acceptance through social innovation. *Energy Policy*, 35(5), 2761–69.

Matanle, P. 2011. The Great East Japan Eartquake, tsunami, and nuclear meltdown: towards the (re)construction of a safe, sustainable, and compassionate society in Japan's shrinking regions. *Local Environment*, 16(9), 823–47.

McDonald, P. 1986. Japan's oil: coping with insecurity. *The World Today*, 42(8/9), 147–49.

McLellan, B.C., Zhang, Q., Utama, N.A., Farzaneh, H. and Ishihara, K.N. 2013. Analysis of Japan's post Fukushima energy strategy. *Energy Strategy Reviews*, 2(2): 190–98.

Moe, E. 2009. Mancur Olson and structural economic change. *Review of International Political Economy*, 16(2), 202–30.

Moe, E. 2012. Vested interests, energy efficiency and renewable energy in Japan. *Energy Policy*, 40(1), 260–73.

Nakata, T. 2002. Analysis of the impacts of nuclear phase-out on energy systems in Japan. *Energy*, 27(4), 363–77.

Namikawa, R. 2003. Take-or-pay under Japanese energy policy. *Energy Policy*, 31(13), 1327–37.

Nemetz, P.N. and Vertinsky, I.B. 1984. Japan and the international market for LNG. *The Columbia Journal of World Business*, 19(1), 70–76.

Nemetz, P.N., Vertinsky, I.B. and Vertinsky, P. 1984/1985. Japan's energy strategy at the crossroads. *Pacific Affairs*, 57(4), 553–76.

Oshima, K., Suzuki, T. and Matsuno, T. 1982. Energy issues and policies in Japan. *Annual Review of Energy*, 7, 87–108.

Pang Er, L. 2010. The Hatoyama administration and Japan's climate change initiatives. *East Asian Policy*, 2(1), 69–77.

Perkins, F.C. 1994. A dynamic analysis of Japanese energy policies. *Energy Policy*, 22(7), 595–607.

Peters, M., Schneider, M., Griesshaber, T. and Hoffman, V.H. 2012. The impact of technology-push and demand-pull policies on technical change – Does the locus of policies matter? *Research Policy*, 41(8), 1296–1308.

Phillips, A. 2013. A dangerous synergy: energy securitization, great power rivalry and strategic stability in the Asian century. *The Pacific Review*, 26(1), 17–38.

Ramana, M.V. 2011. Nuclear power and the public. *Bulletin of Atomic Scientists*, 67(4), 43–51.

Ramseyer, J.M. 2012. Why power companies build nuclear reactors on fault lines: the case of Japan. *Theoretical Inquiries in Law*, 13(2), 457–85.

Rubin, E. 2013. Climate change, technology innovation, and the future of coal. *Cornerstone: The Official Journal of the World Coal Industry*, 1(1), 37–43.

Salameh, M.G. 2001. A third oil crisis? *Survival*, 43(3), 129–44.

Samuels, R.J. 1989. Consuming for production: national security, the domestic economy, and nuclear fuel procurement in Japan. *International Organization*, 43(4), 625–46.

Sano, T. 2011. Improving efficiency of thermal power generation in Japan. *Journal of Power and Energy Systems*, 5(2), 146–60.

Scalise, P.J. 2011c. Looming electricity crisis: three scenarios for economic impact. *The Oriental Economist*, 79(4), 8–9.

Scalise, P.J. 2012c. Not enough land for solar. *The Oriental Economist*, 80(11), 8–11.

Schaede, U. 1995. The Old Boy network and government–business relationships in Japan. *Journal of Japanese Studies*, 21(2), 293–317.

Shadrina, E. 2012. Fukushima fallout: gauging the change in Japanese nuclear energy policy. *International Journal of Disaster Risk Science*, 3(2), 69–83.

Shaoul, R. 2005. An evaluation of Japan's current energy policy in the context of the Azedegan oil field agreement signed in 2004. *Japanese Journal of Political Science*, 6(3), 411–37.

Sheingate, A.D. 2003. Political entrepreneurship, institutional change, and American political development. *Studies in American Political Development*, 17(3), 185–203.

Sinha, N.P. 1974. Japan and the oil crisis. *The World Today*, 30(8), 335–44.

Sovacool, B.K. and Vivoda, V. 2012. A comparison of Chinese, Indian, and Japanese perceptions of energy security. *Asian Survey*, 52(5), 949–69.

Stigler, G.J. 1971. The theory of economic regulation. *Bell Journal of Economics and Management Science*, 2(1), 3–21

Sugawara, S., Inamura, T, Kimura, H. and Madarame, H. 2009. Analysis of relationship between the local governments and the power companies. *Transactions of the Atomic Energy Society of Japan*, 8(2), 154–64.

Surrey, J. 1974. Japan's uncertain energy prospects: the problem of import dependence. *Energy Policy*, 2(3), 204–30.

Swinbanks, D. 1991. Japanese nuclear power: political fallout anticipated. *Nature*, 349(6311), 644.

Takase, K. and Suzuki, T. 2011. The Japanese energy sector: current situation, and future paths. *Energy Policy*, 39(11), 6731–44.

Tanaka, N. 2013. Big bang in Japan's energy policy. *Energy Strategy Reviews*, 1(4): 243–46.

Thomas, S. 2012. What will the Fukushima disaster change? *Energy Policy*, 45: 12–17.

Toichi, T. 1994. LNG development at the turning point and policy issues for Japan. *Energy Policy*, 22(5), 371–77.

Toichi, T. 2003. Energy security in Asia and Japanese policy. *Asia–Pacific Review*, 10(1), 44–51.

Tsurumi, Y. 1975. Japan. *Daedalus*, 104(4): 113–27.

Tsurumi, Y. 1978. The case of Japan: price bargaining and controls on oil products. *Journal of Comparative Economics*, 2(2), 126–43.

Unruh, G.C. 2000. Understanding carbon lock-in. *Energy Policy*, 28(12), 817–30.

Ushiyama, I. 1999. Wind energy activities in Japan. *Renewable Energy*, 16(1–4), 811–16.

Valentine, S.V. 2011. Japanese wind energy development policy: grand plan or group think? *Energy Policy*, 39(11), 6842–54.

Valentine, S.V., Sovacool, B.K. and Matsuura, M. 2011. Empowered? Evaluating Japan's national energy strategy under the DPJ administration. *Energy Policy*, 39(3), 1865–76.

Vivoda, V. 2009. Diversification of oil import sources and energy security: a key strategy or an elusive objective? *Energy Policy*, 37(11), 4615–23.

Vivoda, V. 2010. Evaluating energy security in the Asia–Pacific region: A novel methodological approach. *Energy Policy*, 38(9), 5258–63.

Vivoda, V. 2012. Japan's energy security predicament post-Fukushima. *Energy Policy*, 46, 135–43.

Vivoda, V. 2014. LNG import diversification in Asia. *Energy Strategy Reviews*, 2(3/4), 289–97.

Vivoda, V. and Manicom, J. 2011. Oil import diversification in Northeast Asia: a comparison between China and Japan. *Journal of East Asian Studies*, 11(2), 223–54.

Walsh, J.I. 2006. Policy failure and policy change: British security policy after the Cold War. *Comparative Political Studies*, 39(4), 490–518.

Wang, Q. and Chen, X. 2011. Nuclear accident like Fukushima unlikely in the rest of the world? *Environmental Science & Technology*, 45(23), 9831–32.

Wang, Q. and Chen, X. 2012. Regulatory failures for nuclear safety – the bad example of Japan – implication for the rest of the world. *Renewable and Sustainable Energy Reviews*, 16(5), 2610–17.

Watanabe, C., Wakabayashi, K. and Miyazawa, T. 2000. Industrial dynamism and the creation of a 'virtuous cycle' between R&D, market growth and price reduction. The case of photovoltaic power generation (PV) development in Japan. *Technovation*, 20(6), 299–312.

West, J. 2013. Japanese investment in Australian coal assets through the demise of concessional financing. *Energy Policy*, 52, 513–21.

Williams, R.H. 2013. Coal/biomass coprocessing strategy to enable a thriving coal industry in a carbon-constrained world. *Cornerstone: The Official Journal of the World Coal Industry*, 1(1), 51–59.

Wilson, J.D. 2012. Resource security: a new motivation for free trade agreements in the Asia-Pacific region. *The Pacific Review*, 25(4), 429–53.

Wood, D.A. 2012. A review and outlook for the global LNG trade. *Journal of Natural Gas Science and Engineering*, 9, 16–27.

Wu, G., Liu, L.C. and Wei, Y.-M. 2009. Comparison of China's oil import risk: results based on portfolio theory and a diversification index approach. *Energy Policy*, 37(9), 3557–65.

Yorke, V. 1980. Oil, the Middle East and Japan's search for stability. *International Affairs*, 57(3), 428–48.

Zhang, Q., Ishihara, K.N., McLellan, B.C. and Tezuka, T. 2012a. Scenario analysis on future electricity supply and demand in Japan. *Energy*, 38(1), 376–85.

Zhang, Q., McLellan, B.C., Tezuka, T. and Ishihara, K.N. 2012b. Economic and environmental analysis of power generation expansion in Japan considering Fukushima nuclear accident using a multi-objective optimization model. *Energy*, 44(1), 986–95.

Newspaper, Magazine and Online Journal Articles

ABC News. 2012. Japan's nuclear stocks react after election. 17 December.

Ako, W. 1991. Mishap report won't cool nuclear fears. *The Japan Times*, 26 November.

The Asahi Shimbun. 2011. TEPCO safety management at nuclear plants described as lax. 26 March.

The Asahi Shimbun. 2012. Abe administration to overhaul nuclear, monetary policies. 27 December.

The Australian. 2011. Japanese back nuke free future – poll. 14 June.

BBC News. 2011. Japan PM Naoto Kan urges nuclear-free future. 13 July.

Bloomberg. 2012a. Atomic-free Japan by April roils debate on economic risks. 27 January.

Bloomberg. 2012b. Japan gas binge ties third-biggest economy to one fuel. 24 May.

Buckley, S. 2006. Japan's shaky nuclear record. *BBC News*, 24 March.

Calvo, A. 2012. Can Russia assist Japan in fueling its energy future? *Journal of Energy Security*, 23 July.

CNN. 2011. Anxiety in Japan grows as death toll steadily climbs. 14 March.

Coker, M. 2009. Korean team to build U.A.E. nuclear plants. *The Wall Street Journal*, 28 December.

Cooper, M. 2013. Japanese power, Australian coal producers to resume contract price talks. *Platts*, 8 April.

Cross, M. 1988. Japan's nuclear industry tries to rescue its image. *New Scientist*, 24 March.

Dale, C. 2011. Energy Angst: Japan's Post-Tsunami Power Crisis. *CBC News*, 30 May.

The Denki Shimbun. 2013. Chubu Electric Power, Korea Gas Corporation cooperate in LNG procurement. 18 January.

DeWit, A. 2011. Fallout from the Fukushima shock: Japan's emerging energy policy. *The Asia–Pacific Journal*, 7 November.

DeWit, A. 2012a. Megasolar Japan: the prospects for green alternatives to nuclear power. *The Asia–Pacific Journal*, 23 January.

DeWit, A. 2012b. Japan's remarkable renewable energy drive – after Fukushima. *The Asia–Pacific Journal*, 11 March.

Dickie, M. 2012. Noda's nuclear phase-out is not final word. *Financial Times*, 16 September.

The Economist. 1990. 14 April.

The Economist. 2009. Raising the stakes. 8 April.

The Economist. 2011. Powerful savings. 27 April.

The Economist. 2012a. Japan's trading houses: resourceful and energetic. 12 May.

The Economist, 2012b. LNG: a liquid market. 14 June.

The Economist. 2012c. Japan's energy security: foot on the gas. 22 September.

Energyboom. 2009. Japanese firms target foreign wind growth. 4 August.

Engler, D. 2008. Is Japan a leader in combating global warming? *The Asia–Pacific Journal*, 5 May.

Ferguson, C.D. 2011. The need for a resilient energy policy. *Bulletin of the Atomic Scientists*, 16 March.

Grimond, J. 2002. A survey of Japan. *The Economist*, 18 April.

Harlan, C. 2011. Renewable energy sees its chance in Japan's electricity market. *The Washington Post*, 1 October.

Hosoe, T. 2012a. LNG's role in post-Fukushima Japanese energy policy. *Oil & Gas Journal*, 2 April.

Iida, T. and DeWit, A. 2009. Is Hatoyama reckless or realistic? Making the Case for a 25% cut in Japanese greenhouse gases. *The Asia–Pacific Journal*, 21 September.

Inajima, T. 2012. Japan utilities losses reach $6 billion after Fukushima crisis. *Bloomberg*, 17 January.

Ito, M. 2012. Zero nuke policy: behind the no-nuclear option. *The Japan Times*, 30 October.

The Japan Times. 2009. Rhetorical Hatoyama opens Diet. 27 October.

The Japan Times. 2012. A worrisome trade deficit. 31 January.

Johnston, E. 2011. Son's quest for sun, wind has nuclear interests wary. *The Japan Times*, 12 July.

Kanie, N. 2011. Time to reform Japan's ministries on energy and climate. *The Asahi Shimbun*, 12 October.

Kim, V. 2011. Japan damage could reach $235 billion, World Bank estimates. *Los Angeles Times*, 21 March.

Kingston, J. 2011. Ousting Kan Naoto: the politics of nuclear crisis and renewable energy in Japan. *The Asia–Pacific Journal*, 26 September.

Kingston, J. 2012b. Mismanaging risk and the Fukushima nuclear crisis. *The Asia–Pacific Journal*, 19 March.

Kingston, J. 2012c. Japan's nuclear village. *The Asia–Pacific Journal*, 10 September.

Kingston, J. 2012d. Power politics: Japan's resilient nuclear village. *Japan Focus*, 4 November.

Kitazume, T. 2012. Japan–U.S. seminar: U.S. needs Japan to remain nuclear, expert says. *The Japan Times*, 3 November.

Kojina, R. 2013. Oe: state secrets law endangers freedoms in Japan. *The Asahi Shimbun*, 11 December.

Masaki, H. 2006. Japan's new energy strategy. *Asia Times Online,* 13 January.

McCurry, J. 2013. Japan whistleblowers face crackdown under proposed state secrets law. *The Guardian*, 6 December.

Muramatsu, S., Komori, A. and Hirai, Y., 2011. Kan bent on taking down utility interests with him. *Asahi Shimbun*, 29 June.

Natural Gas Asia. 2011. Increased global LNG demand forecast. 30 August.

Nelder, C. 2013. Are methane hydrates really going to change geopolitics? *The Atlantic*, 2 May.

NPR. 2011. Millions of stricken Japanese lack water, food, heat. 14 March.

Onishi, N. and Belson, K. 2011. Culture of complicity tied to stricken nuclear plant. *The New York Times*, 26 April.

Onitsuka, H. 2012. Hooked on nuclear power: Japanese state–local relations and the vicious cycle of nuclear dependence. *The Asia–Pacific Journal*, 16 January.

PIW (Petroleum Intelligence Weekly). 2001. JNOC's demise leaves Mideast assets in limbo. 6 August.

Pritchard, J., Thibodeaux, T., Sambriski, B. and Reed, J. 2011. Revolving door connects Japanese nuke industry, sympathetic government regulators. *Washington Post*, 1 May.

Reuters. 2013. Japan may restart reactors in a year. 10 July.

Sawa, T. 2013. A nagging Japanese riddle: who's 'left' and who's 'right'? *The Japan Times*, 21 January.

Scalise, P.J. 2012b. Japanese energy policy after Fukushima. *Policy Innovations*.

Sieg, L. and Takenaka, K. 2013. Japan secrecy act stirs fears about press freedom, right to know. *Reuters*, 24 October.

Soble, J. 2012. Tepco raises electricity price by up to 18%. *The Financial Times*, 17 January.

Son, M. and DeWit, A. 2011. Creating a solar belt in East Japan. *The Asia–Pacific Journal*, 19 September.

Stewart, D.T. and Wilczewski, W. 2009. How Japan became an efficiency superpower: lessons for U.S. energy policy under Obama. *Policy Innovations*.

Tabuchi, H. 2011. Japan leader to keep nuclear phase-out. *The New York Times*, 2 September.

Tabuchi, H., Onishi, N. and Belson, K. 2011. Japan extended reactor's life, despite warning. *The New York Times*, 21 March.

Tolliday, S. 2012. Crumbling dream: Japan's nuclear quest, 1954–2011. *Business and Economic History*, vol. 10.

Tsukimori, O. 2012. China overtakes Japan as world's top coal importer. *Reuters*, 26 January.

Tsukimori, O. and Tan, F. 2013. Japan turns to coal as yen drives up energy costs. *Reuters*, 28 March.

Tsukimori, O. and Maeda, R. 2013. Coal exporters jostle for opportunity as Japan ramps up use. *Reuters*, 1 May.

Wakamatsu, J. 2011. 35 prefectures to join Softbank's 'natural energy council'. *The Asahi Shimbun*, 25 June.

Wallace, R. 2011. Japan's energy loss, our LNG gain. *The Australian*, 20 September.
World Nuclear News. 2010. Japanese nuclear exports consortium launches. 26 October.
World Nuclear News. 2012. Trade figures reveal cost of Japan's nuclear shutdown. 25 January.

Reports, Conference Papers and Think-tank Publications

Aldrich, D.P. 2012a. Post-crisis Japanese nuclear policy: from top-down directives to bottom-up activism. East–West Center, AsiaPacific Issues, no. 103, January.
Andrews-Speed, P., Liao, X. and Dannreuther, R. 2002. The strategic implications of China's energy needs. Adelphi Paper 346, Oxford University Press.
Andrews-Speed, P. 2012. Do overseas investments by national oil companies enhance energy security at home? A view from Asia. National Bureau of Research (NBR), NBR Special Report, no. 41, September.
Barrett, B. 2011. Can nuclear power save Japan from peak oil? United Nations University, Tokyo.
Bhattacharya, A., Janardhanan, N.K. and Kuramochi, T. 2012. Balancing Japan's energy and climate goals: exporting post-Fukushima energy supply options. Report of the Disaster Study Project conducted jointly by Economy and Environment Group (EE) and Climate Change Group (CC), Institute for Global Environmental Strategies (IGES), Japan.
BP. 2013. *Statistical review of world energy*. London: BP.
Christoffels, J.-H. 2007. Getting to grips again with dependency: Japan's energy strategy. Clingendael, Netherlands Institute of International Relations, August.
EDMC (Energy Data and Modelling Center), Institute of Energy Economics, Japan. 2010. EDMC Handbook of Energy & Economic Statistics in Japan. The Energy Conservation Center, Tokyo, Japan.
EIA (Energy Information Administration), US Department of Energy. 2011a. Japan energy data, statistics and analysis. March.
EIA (Energy Information Administration), US Department of Energy. 2011b. Australia: country analysis brief. 28 October.
EIA (Energy Information Administration), US Department of Energy. 2012a. Effect of increased natural gas exports on domestic energy markets as requested by the Office of Fossil Energy, January.
EIA (Energy Information Administration), US Department of Energy. 2012b. Japan: country analysis brief. 4 June.
Energy Charter Secretariat. 2010. Putting a price on energy: international coal pricing.
EPIA. 2010. 2015: Global Market Outlook for Photovoltaics until 2015. European Photovoltaic Industry Association Renewable Energy House, Brussels.
Ernst and Young. 2009. Renewable energy country attractiveness indices. Issue 22, August.

Evans, P.C. 2006. Japan. The Brookings Foreign Policy Studies, Energy Security Series, December.

FEPC (The Federation of Electric Power Companies of Japan). 2012. Electricity review Japan.

Fesharaki, F. and Hosoe, T. 2011. The Fukushima crisis and the future of Japan's power industry. *Asia Pacific Bulletin*, No. 106, East–West Center, 12 April.

GIIGNL (International Group of Liquefied Natural Gas Exporters). 2013. *The LNG industry*

in 2012. Paris: GIIGNL.

Hanser, P.Q. 2012. Impact of U.S. LNG on international gas prices. Paper presented to EIA International Natural Gas Workshop, 23 August.

Hartley, P. 2000. Reform of the electricity supply industry in Japan. Japanese Energy Security and Changing Global Energy Markets: An Analysis of Northeast Asian Energy Cooperation and Japan's Evolving Leadership Role in the Region. Prepared in conjunction with an energy study sponsored by The Center for International Political Economy and The James A. Baker III Institute for Public Policy, Rice University, May.

Herberg, M.E. 2007. The rise of Asia's national oil companies. National Bureau of Asian Research (NBR), NBR Special Report, no. 14, December.

Hosoe, T. 2012b. Asia's post-Fukushima market for liquefied natural gas: a special focus on Japan. National Bureau of Research (NBR), NBR Special Report, no. 41, September.

IAE (Institute of Applied Energy, Japan). 2009. '発電方式別の発電コストの比較' ['Cost comparison of different power generation systems'], 7 July.

IEA (International Energy Agency). 2006. Trends in photovoltaic applications: survey report of selected IEA countries between 1992 and 2005. Report IEA-PVPS T1–15, August.

IEA (International Energy Agency). 2010a. Global renewable energy policies and measures database.

IEA (International Energy Agency). 2010b. Trends in photovoltaic applications. Report IEA-PVPS T1–19:2010.

IEA (International Energy Agency). 2011b. Energy perspective after the accident at Fukushima.

IEA (International Energy Agency). 2013a. Monthly energy statistics. October 2012.

IEA (International Energy Agency). 2013b. Closing oil stock levels in days of net imports. February.

IEEJ (Institute of Energy Economics, Japan). 2010. Basic Energy Plan drafted for 2010 revision. Japan Energy Brief, no. 7, May.

IEEJ (Institute of Energy Economics, Japan). 2011a. Analysis of electricity supply and demand through FY2012 regarding restart of nuclear power plants. Special bulletin, 13 June.

IEEJ (Institute of Energy Economics, Japan). 2011b. Japan energy brief, No. 14, July.

IEEJ (Institute of Energy Economics, Japan). 2011c. Japan energy brief, No. 15, September.
IEEJ (Institute of Energy Economics, Japan). 2011d. Japan energy brief, No. 16, November.
IEEJ (Institute of Energy Economics, Japan). 2012a. Japan energy brief, No. 17, January.
IEEJ (Institute of Energy Economics, Japan). 2012b. Japan energy brief, No. 18, March.
IEEJ (Institute of Energy Economics, Japan). 2012c. IEEJ e-newsletter, No. 1, 13 April.
IEEJ (Institute of Energy Economics, Japan). 2012d. IEEJ e-newsletter, No. 4, 20 July.
IEEJ (Institute of Energy Economics, Japan). 2012e. IEEJ e-newsletter, No. 5, 17 August.
IEEJ (Institute of Energy Economics, Japan). 2012f. IEEJ e-newsletter, No. 6, 19 September.
IEEJ (Institute of Energy Economics, Japan). 2012g. IEEJ e-newsletter, No. 7, 19 October.
IEEJ (Institute of Energy Economics, Japan). 2012h. IEEJ e-newsletter, No. 8, 19 November.
IEEJ (Institute of Energy Economics, Japan). 2012i. IEEJ e-newsletter, No. 9, 18 December.
IEEJ (Institute of Energy Economics, Japan). 2013a. IEEJ e-newsletter, No. 10, 21 January.
IEEJ (Institute of Energy Economics, Japan). 2013b. IEEJ e-newsletter, No. 11, 15 February.
IEEJ (Institute of Energy Economics, Japan). 2013c. IEEJ e-newsletter, No. 12, 15 March.
IEEJ (Institute of Energy Economics, Japan). 2013d. IEEJ e-newsletter, No. 14, 16 May.
IEEJ (Institute of Energy Economics, Japan). 2013e. IEEJ e-newsletter, No. 17, 12 July.
Iida, T. 2010. Policy and politics of renewable energy in Japan. Paper presented at Rikkyo University International Symposium: The Energy Revolution and the Comparative Political Economy of the FIT.
Iijima, M. 2012. Japan focuses on next generation biofuels. USDA Foreign Agricultural Service, Global Agricultural Information Network, 29 June.
Itakura, K. 2011. The economic consequences of shifting away from nuclear energy. Economic Research Institute for ASEAN and East Asia (ERIA), Policy Brief, no. 2011–04, December.
JPC-SED (Japan Productivity Center for Socio-Economic Development), Energy and Environmental Policy Section, Special Committee of Energy Issues. 2000. *Enerugii sekyuriti no kakuritsu to 21 seiki no enerugii seisaku no arikata – yûshikisha ankeeto chôsa ni motozuite* [The establishment of energy security

and ideal energy policies in the 21st century – based on a survey of learned persons]. Tokyo, 30 March.
Jupesta, J. and Suwa, A. 2012. Sustainable energy policy in Japan post Fukushima. International Association for Energy Economics, IAEE Bulletin 2012 (4th Quarter): 23–26.
Kasai, Y. 2012. Nuclear energy is indispensable for Japan's future. The Association of Japanese Institutes of Strategic Studies, AJISS-Commentary No. 165, 13 November.
Kaufmann, D. 2011. Preventing nuclear meltdown: assessing regulatory failure in Japan and the United States. Brookings, 1 April.
Kimura, O. and Suzuki, T., 2006. 30 years of solar energy development in Japan. Paper presented at Berlin Conference on the Human Dimensions of Global Environmental Change.
Kotler, M.L., and Hillman, I.T. 2000. Japanese nuclear energy policy and public opinion. Japanese Energy Security and Changing Global Energy Markets: An Analysis of Northeast Asian Energy Cooperation and Japan's Evolving Leadership Role in the Region. Prepared in conjunction with an energy study sponsored by The Center for International Political Economy and The James A. Baker III Institute for Public Policy, Rice University, May.
Koyama, K. 2012. A new development for Japan's procurement of U.S. LNG. IEEJ Special Bulletin, 10 August.
Kral, A. 2000. Are Japan's energy security ambitions misguided? Nuclear policy in Japan. Event Summary, Woodrow Wilson International Center for Scholars, 29 February.
Leiby, P.N. 2007. Estimating the energy security benefits of reduced US oil imports. Oak Ridge National Laboratory, 23 July.
Luta, A. 2010. Ghosts of Crises Past. Working Papers 66:2010, Tokyo Institute of Technology.
Martin, W.F. 2011. Achieving Japanese energy security in a post-Fukushima world. Partnership for Recovery, Center for Strategic and International Studies, 3 August.
Matsuo, Y., Nagatomi, Y. and Murakami, T. 2011. Thermal and nuclear generations cost estimates using corporate financial statements. Institute for Energy Economics, Japan, October.
Medlock, K.B. and Hartley, P. 2004. The role of nuclear power in enhancing Japan's energy security. Paper prepared in conjunction with an energy study sponsored by Tokyo Electric Power Company Inc. and The James Baker III Institute for Public Policy, Rice University, March.
Meltzer, J. 2011. After Fukushima: what's next for Japan's energy and climate change policy?. Global Economy and Development at Brookings, 7 September.
Miyamoto, A., Ishiguro, C. and Nakamura, M. 2012. A realistic perspective on Japan's LNG demand after Fukushima. The Oxford Institute for Energy Studies, NG 62, June.

Miyamoto, A., Ishiguro, C. and Yamada, T. 2009. Irrational LNG pricing impedes development of Asian natural gas markets: a perspective on market value. Paper presented at the 24th World Gas Conference, Buenos Aires, Argentina, 5–9 October.

Nakano, J. 2011. Japan's energy supply and security since the March 11 earthquake. Center for Strategic & International Studies, Commentary, 23 March.

Oyama, K. 2000. Trends and prospects of deregulation in Japan. Japanese Energy Security and Changing Global Energy Markets: An Analysis of Northeast Asian Energy Cooperation and Japan's Evolving Leadership Role in the Region. Prepared in conjunction with an energy study sponsored by The Center for International Political Economy and The James A. Baker III Institute for Public Policy, Rice University, May.

Roney, J.M. 2010. Solar Cell Production Climbs to another Record in 2009. Earth Policy Institute.

Sagawa, A. 2012. Coal for power generation: supply and demand following the Great East Japan Earthquake. IEEJ, August.

Scalise, P.J. 2011a. Rethinking national energy policy: Japan's electricity crisis, nuclear power, and the fate of TEPCO. Temple University, Institute of Contemporary Asian Studies, 27 May.

Smith, S.A. 2012. Japan's Iran sanctions dilemma. Council on Foreign Relations, Asia Unbound, 31 January.

Soligo, R. 2001. New energy technologies in the natural gas sectors: a policy framework for Japan. Prepared in conjunction with an energy study sponsored by The Center for International Political Economy and The James A. Baker III Institute for Public Policy, Rice University, November.

Standard & Poor's. 2012. What's behind the boom in global liquefied natural gas development? Global Credit Portal, 20 April.

Stevens, P. 2010. The 'shale gas revolution': hype and reality. A Chatham House Report, September.

Sugino, H. and Akeno, T. 2010. 2010 country update for Japan. World Geothermal Congress, Bali, Indonesia, 25–29 April.

Suzuki, T. 2000a. Nuclear power generation and energy security: The challenges and possibilities of regional cooperation. Japanese Energy Security and Changing Global Energy Markets: An Analysis of Northeast Asian Energy Cooperation and Japan's Evolving Leadership Role in the Region. Prepared in conjunction with an energy study sponsored by The Center for International Political Economy and The James A. Baker III Institute for Public Policy, Rice University, May.

Suzuki, T. 2000b. Energy security and the role of nuclear power in Japan. Nautilus Institute.

Tanaka, N. 2012. The Middle East situation and post-Fukushima energy strategy: a compound crisis for Japan? Paper presented at the Japan Institute of International Affairs, 9 October.

Toichi, T. 2006. International energy security and Japan's strategy. Paper presented at the conference on India's energy security, jointly organized by TERI and the Konrad Adenauer Foundation, Goa, India, 29–30 September.
Toichi, T. 2008. Japan's energy challenges and the rise of gas. Paper presented at Gastech Conference, Bangkok, 10–13 March.
World Bank. 2013. Japan: data. Available at: http://data.worldbank.org/country/japan [accessed 23 May 2013].
World Nuclear Association. 2012. Nuclear power in Japan. 22 October.
WWEA (World Wind Energy Association), 2011. World wind energy report 2010.
Yamashita, Y. 2012. Challenges for energy policy in Japan after the Great Earthquake. Saving Electricity in a Hurry, Beijing, 23 February.
Yoshizaki, K., Sato, N., Fukagawa, H., Sugiyama, H., Takagi, G. and Jono, T. 2009. Utilization of underground gas storage (UGS) in Japan. 24th World Gas Conference, Buenos Aires, Argentina, 5–9 October.

Policy and Government Documents

ANRE (Agency for Natural Resources and Energy, Japan). 2006. Energy in Japan 2006: status and policies.
Government of Japan. 2005. Kyoto Protocol target achievement plan.
JAEC (Japan Atomic Energy Commission). 1957. The long-term plan for the research, development and use of nuclear power. 6 September.
JAEC (Japan Atomic Energy Commission). 2009. White paper on nuclear energy 2008 (Summary). March.
METI (Ministry of Economy, Trade and Industry, Japan). 2006a. New national energy strategy (Digest). May.
METI (Ministry of Economy, Trade and Industry, Japan) 2006b. The challenges and directions for nuclear energy policy in Japan: Japan's nuclear energy national plan. Nuclear Energy Policy Planning Division, December.
METI (Ministry of Economy, Trade and Industry, Japan). 2010a. The strategic energy plan of Japan.
METI (Ministry of Economy, Trade and Industry, Japan). 2010b. The international situation and the main issues about the nuclear power.
METI (Ministry of Economy, Trade and Industry, Japan). 2011a. 2011 annual report on energy (Energy White Paper 2011). October.
METI (Ministry of Economy, Trade and Industry, Japan). 2011b. Feed-in tariff scheme for renewable energy resources and energy.
METI (Ministry of Economy, Trade and Industry, Japan). 2011c. Preparations for the introduction of the feed-in tariff scheme for renewable energy.
METI (Ministry of Economy, Trade and Industry, Japan). 2011d. Act on purchase of renewable energy sourced electricity by electric utilities.

METI (Ministry of Economy, Trade and Industry, Japan). 2012a. Reform of Japan's nuclear safety regulation.

METI (Ministry of Economy, Trade and Industry, Japan). 2012b. The current state of LNG supply in Japan.

MoF (Ministry of Finance, Government of Japan). 2013. Trade statistics of Japan. Available at: http://www.customs.go.jp/toukei/info/index_e.htm [accessed: 20 May 2013].

MoFA (Ministry of Foreign Affairs, Japan). 2004. Strategy and approaches of Japan's energy diplomacy. April.

MoFA (Ministry of Foreign Affairs, Japan). 2008. Japan's official development assistance white paper 2008.

National Police Agency of Japan. 2013. Damage situation and police countermeasures associated with 2011 Tohoku district – off the Pacific Ocean earthquake. 6 February.

State Council (People's Republic of China). 2007. China's energy conditions and policies. Information Office of the State Council.

Index